The Nautical Institute on Command

Third edition

The Nautical Institute on Command
Third edition

Published by The Nautical Institute
202 Lambeth Road, London SE1 7LQ, UK
Tel: +44 (0)20 7928 1351   Fax: +44 (0)20 7401 2817   Web: www.nautinst.org
© The Nautical Institute 2015

All rights reserved. No part of this publication may be reproduced, stored in a retrieval system, or transmitted in any form or by any means, electronic, mechanical, photocopying, recording or otherwise, without the prior written consent of the publisher, except for quotation of brief passages in reviews.

Although great care has been taken with the writing of the book and the production of the volume, neither The Nautical Institute nor the contributors can accept any responsibility for errors and omissions or their consequences.

This book has been prepared to address the subject of Command. This should not, however, be taken to mean that this document deals comprehensively with all of the concerns that will need to be addressed or even, where a particular matter is addressed, that this document sets out the only definitive view for all situations. The opinions expressed are those of the contributors only and are not necessarily to be taken as the policies or views of any organisation with which they have any connection.

Readers of *The Nautical Institute on Command* are advised to make themselves aware of any applicable local, national or international legislation or administrative requirements or advice which may affect decisions taken on board.

Book Editor Margaret Freeth
Typesetting and layout by Phil McAllister Design
Printed in the UK by Geerings Print Ltd

ISBN 978 1 906915 21 6

Acknowledgements

A book of this scope and depth was made possible by the contributions of many individuals and The Nautical Institute would like to thank all who were involved as authors and peer reviewers and those who gave advice and support.

Special thanks are due to the following:

The Technical Editors, Captain Trevor Bailey FNI, Chairman of the Technical Committee, and Captain John Dickinson FNI, who is leading the revision of the Institute's Command Scheme. They drew up the voyage plan, recruited the authors and technically reviewed their contributions, and ensured that the whole is greater than the sum of its parts.

All the authors for their generosity in sharing their expertise, particularly those who also peer reviewed other articles.

Our peer reviewers: Captain Amadeu Alburquerque MNI; Capt Satish P Anand FIIISLA HonFIIMS RMC MNI FICA; Shernelle Cox; Captain Flavian D'Souza MNI; Captain Krish Krishnamurthi FNI; Captain George Livingstone FNI; Murray Milligan; Mark Revill.

The SE Australia Branch which took on the subject of running meetings and taking minutes as a project, of which the article in this book is only a summary.

Theresa Nelson and all the members of the Command Scheme revision working group.

Captain Nikos Aslanis of Tsakos Columbia Shipmanagement.

Foreword

By **Rear Admiral Nick Lambert**
CMarTech FIMarEST AFNI

Many, many congratulations. You've done it! Years of hard work, seagoing, academia and dedication have paid off and you're in command! No matter what lies ahead, no one and nothing can deprive you of this moment and of your achievement. You've joined the ranks of several centuries of illustrious forebears to confront the challenges of the seas and oceans in command of your ship. So, after a few justifiably self-congratulatory moments inspecting the Captain's suite, revelling in the sensation of the Captain's chair on the bridge and absorbing the scale of your responsibility, what next?

Firstly, be in no doubt that the ship is yours. She is of course owned by others but, while you are in command, she is for all intents and purposes yours. You set the style, the character, and the reputation of your ship, your people and your operations more so, I would argue, than in any other walk of life. Notwithstanding the relentless pace of technology, today's manned ships are geographically independent units confronting a notoriously unpredictable environment frequently beyond reach of shore support and often outside the comprehension of most people ashore – and you're in charge of one.

Secondly, the operation of your ship reflects your personal professionalism and that of your people. While shipping is often regarded as 'out of sight, out of mind' in reality that is rarely the case. Modern satellite and terrestrial networks enable almost constant communication such that agents, shipping operators and companies, ports, harbours, vessel traffic services and numerous other organisations abroad and at home (and of course families and friends) can interact with you and your team. The professionalism of your response will be assessed and interpreted far afield and conclusions about your ship, positive or otherwise, will be drawn. Equally your people will subliminally adopt and mirror your professional style.

Thirdly, when the chips are down, it's your sense of purpose and your commitment to your mission, your company and your people that will ensure that you with your team deliver your objectives – nobody else can or will. Responsible for a myriad of tasks, your leadership will ensure the availability of systems and equipment, the making of an ETA or ETD and the satisfaction of your company and clients. Your drive is what counts.

Finally (and by no means least), there's the people. It's no mistake that the word people appears throughout this introduction because, above all, the crew of your ship are your people. You set the tone for the quality of their life on board; you set the standards of behaviour and welcome; you are the arbiter of their lives at sea and they will subconsciously adopt your style. In fact you have a unique opportunity to influence the lives and careers of your juniors; people always remember Captains and how they led. It's an awesome but intensely gratifying responsibility and experience.

Now this is at first sight a daunting list, however, when you've had time to reflect, you'll realise that it's not insurmountable. After years of training you're prepared for this. At last you have the authority, freedom and joy of stamping your personality and professionalism on another generation of seafarers, you have a unique opportunity to make a personal contribution to the lives of those who work for you, to the company that employs you and, without being overly dramatic, to contribute to the efficient global ebb and flow of raw materials and products that underpin the world's economy. Enjoy every single second of your time, strive for the best and ensure that you and your people have fun. Good luck!!

Contents

Introduction by Captain Trevor Bailey 1

SECTION 1: THIS IS THE JOB

Major conventions by Captain Robert Kieran 5
Managing people on board by Captain Kuba Szymanski 8
Managing multicultural crews by Captain André L Le Goubin 9
Master's standing orders by Captain Nick Nash 10

RELATIONSHIPS WITH STAKEHOLDERS

Who are the stakeholders? by Graham Cowling 16
Relationships with flag state by James Parkhouse 19
Relationships with class by Peter Hamer 21
Relationships with charterers and owners by Captain Ghulam Hussain 23
Relationships with pilots by Captain Sean Bolt 26
Relationships with port state control by Captain Neil Forde 28
Relationships with insurers by Chris Adams 31
The Master's responsibilities in law by Captain Richard Springthorpe 33

SECTION 2: OPERATIONAL ASPECTS

Handover by Captain Ozan Dermen 39
ISM expectations by Kevin Slade 42
Fatigue management by Michelle Grech 44

STATE OF THE SHIP

State of the ship: cargo by Ian MacLean 47
State of the ship: certification by Kenny Crawford 50
State of the ship: surveys by Walter Vervloesem 52

SECTION 3: DAY TO DAY MANAGEMENT – OPERATIONS

Shiphandling by Captain Trevor Bailey 57
Safe working practices by Captain Nicholas Cooper 59
Safe management and delivery of cargo by Captain Richard Brough 61
Effective drills by Captain Sarabjit Butalia 63
Inspection and maintenance by Captain Sanjay Bhasin 66
Ice by Captain Duke Snider 69
Writing reports by Lucy Budd 73
Running meetings and taking minutes by Jillian Carson-Jackson and Christine Dickinson .. 76

SECTION 4: DAY TO DAY MANAGEMENT – PERSONNEL MANAGEMENT

Relationships on board by Dr Captain François Laffoucrière 79
MLC pastoral care by Reverend Canon Ken Peters 81
Health, well-being and hygiene by Dr Toby Abaya 84
Food and catering by Tapan Kumar 89
Onboard training and development by Captain André L Le Goubin 91
Discrimination by Captain Wendy Maughan 94
Appraisals by Captain Sriram Rajagopal 96
Discipline by Iain Macleod 98
Managing conflict at sea by Alison Williams 101
The basics of negotiation by Captain Kuba Szymanski 104

SECTION 5: WHAT IF?

Accidents on board by Captain Paul Drouin 107
Medevac – helicopter arrangements by Captain Eric Patten 108
Medevac – onboard arrangements by Captain Pushkar Gadam 112
Dealing with an oil spill by Captain Alex van Wijngaarden 114
Emergency response by Captain Alex van Wijngaarden 116
Criminalisation by Captain James Robinson 119
Crime at sea by Steven Jones 121
Dealing with death on board – the emotional impact by Lynda Bailey 123
Dealing with death on board – the practicalities by Captain Alexander Sagaydak 125
Dealing with the media – professional and social by Steven Jones 127

Contributors 131

THE NAUTICAL INSTITUTE

Introduction

By Captain Trevor Bailey, Technical Editor, *The Nautical Institute on Command*

The Nautical Institute aims to demonstrate and to represent the highest possible standards in the nautical profession and it is only right that the Institute should ensure that its advice to those in the highest positions of responsibility – those in command of sea-going ships – represents those highest standards. *The Nautical Institute on Command* has been, and should remain, our flagship publication.

In 1986, The Nautical Institute published the first edition of *Command* containing advice to those aspiring to command. Recognising the dynamic nature of our industry, this was revised in 2000.

Since then, our industry and our membership have continued to evolve and the time has come for another review – not least recognising that technology has significantly changed the way in which we work over the intervening period. We have also recognised that The Nautical Institute is a truly international membership organisation that covers many different areas of operation of our diverse industry across the globe, including an increase from the military maritime world. We must also not lose sight of the fact that the old traditions of a male-only society no longer prevail and that our female members have a lot to offer our industry.

In this third edition, we have followed the same format as before, providing essays on a wide variety of subjects that are relevant to the exercise of command and which may not have been fully covered during the training years. Our authors have volunteered from many of our branches, thus allowing us to present a global view on command. Each of them was given the same general brief – you have 15 minutes over a cup of coffee to spend with the new Master before they go off in the taxi to join their ship: how can you best distil your advice into that time frame?

The result is a volume of consolidated advice from experts in their field.

It would be unfair to single out any particular contributing author at the expense of the others but I would like to use just one quote from one of the articles:

It is a temptation for young Masters to continue to do their former job instead of the new one. You need to put some distance between the two roles.

This is probably one of the best pieces of advice in this book and one which I encourage you all to consider when you finally sit in the Master's chair. When I made a belated decision that maybe a career at sea would suit me and I had joined my first ship, I realised that the Master's job was one that I could aspire to. In the fullness of time, when I was the Master suddenly I realised the buck stopped with me! Where was the person to turn to, whose advice could I seek? Could I do the job? There were many questions going through my mind. I wish I had had that advice in my head at the time.

Introduction
The Nautical Institute on Command

I think it would be fair to say that, for a large number of The Nautical Institute's younger sea-going members, aspiration for command is a very strong driver in their career paths; command of a ship will represent a pinnacle of achievement and a tremendous sense of self-satisfaction. For me, command of a ship is an amalgamation of all the skills and knowledge that I acquired during my career; the professional and technical skills, added to the application of soft skills such as personnel management plus a degree of commercial awareness. Masters may not always be able to have a direct influence on the commercial success of the venture of taking ships from one port to another – but they will almost certainly have an indirect influence if they don't manage to command their ships effectively.

I am sure that all of our members will have sailed with Masters and Commanding Officers who were inspirational – "if I ever get that far I want to be like them". Then there were those whom we would not wish to emulate – "surely, I won't be like them?" Then there were some who were totally forgettable. Whoever they were and whatever influence they had upon you, you will probably have learnt something from them. In this book we have tried to distil those learning experiences from many of our colleagues and to pass them on for your benefit.

In putting the book together, we have tried to develop our own voyage plan to lead readers through the basics of command, looking at legislation and how it will impact on the day job. From there, the day to day relationships with the many and varied players that Masters will have to deal with are explained by representatives of these various parties, including flag, class and port state inspectors.

The day to day operation of the ship is addressed in Sections 2, 3 and 4, starting with taking over of command and moving on to navigation and cargo management, surveys and inspections, before detailing some of the soft skills that new Masters will need.

We decided that Section 5 should have the working title What if? Emergency management should have been a part of all Masters' training throughout their careers but, unless you have experienced the real thing, training cannot always provide you with the information that you need to effectively manage what you may encounter.

Throughout we have attempted to collate relevant and experienced advice. It has been very interesting for me to have been a part of this process; although I have over 40 years' experience in our wonderful industry, I like to think that I am never too old to learn and in reading many of the submissions we received for publication, I was reminded of experiences and incidents that I had forgotten. Thus, this book is not just for the aspiring officer or for the new Master – it is just as relevant to the serving Master of several years standing.

Please remember that, since the last edition, The Nautical Institute has increased the range and quality of its publications: this is no longer a standalone publication but it should be read in conjunction with many of our other publications; we have tried to provide cross-references to those publications at the end of any articles where this is relevant.

To everyone who has contributed to this book, whether by writing an article or by peer review, by email contributions or with telephone advice, please accept my sincere and heartfelt thanks. I am confident that we can say that we have met our goals and that this book distils the best of advice to the new Shipmaster.

I would also like to express my deepest admiration and thanks to the Publications Team at NIHQ – your patience, understanding and commitment to this project, along with your amazing interpersonal skills in dealing with all of us with such good grace and professionalism, are second to none and you can pride yourselves on a job well done. In Royal Navy speak Bravo Zulu!

Trevor Bailey
Ynys Môn
May 2015

Section 1

This is the job

Major conventions

By Captain Robert Kieran

Masters have a vital role in overseeing the implementation of mandatory rules and regulations for the normal safe operation of ships and in emergencies.

There are many international conventions that provide a legal framework for ships to carry passengers and cargo, for compensation for loss of life and for damage caused by ships. Many of these international conventions allow shipowners and operators to limit their financial liability provided they operate their ships to an acceptable standard. Without this ability to limit liability, ships and their passengers and cargoes could not be insured at a reasonable cost.

A ship's flag state controls many aspects of the construction, crewing and operation of the ship. The IMO is the primary source of a great many statutory conventions, codes and guidelines but has no powers of its own to implement and enforce the regulations it produces. It relies on its 170 member states to incorporate IMO regulations into their national law. Each member state's flag administration ensures the compliance of its national fleet through surveys and inspections. Maritime states also have a responsibility to ensure the compliance of foreign ships calling at their ports through port state control (PSC) inspections.

The Master is expected to be well informed and in charge of the implementation onboard of the requirements under the laws and regulations that affect the ship's operation. Local laws and regulations that impinge on the ship may change as it moves to different sea areas and ports, and the Master must be alert to these changes and make sure that the vessel, its officers and its crew are adequately prepared for each changing circumstance. These may include physical conditions such as restrictions on navigation for safety reasons, mandatory environmental laws or regulations of a port or coastal state, and mandatory coastal state reporting requirements when transiting a strait.

SOLAS, STCW, MARPOL and MLC are the primary sources of international maritime regulation affecting the operation of ships and the Master must be aware of the provisions in them that are relevant to the ship under command. The ISM Code (Chapter IX of the SOLAS Convention) requires a safety management system (SMS) to contain procedures that will enable the ship's staff to implement the mandatory rules and regulations set out in the conventions. A major objective of the ISM Code is to support and encourage the development of a safety culture in shipping. The Master must ensure that all personnel onboard carry out their duties in the SMS and that they understand the extent of the Master's authority and responsibility in the SMS, particularly in an emergency.

The main objective of the **SOLAS** Convention is to specify minimum standards for the construction, equipment and safe operation of ships. Flag states are responsible for ensuring that ships under their flag comply with SOLAS requirements and that specified certificates such as Load Line, ISM Code, Safe Manning and LSA are up to date.

The SMS is your main reference resource and guide to the mandatory rules, regulations and guidelines and to company requirements and practice as they relate to the ship under command. It is a requirement of the ISM Code that the SMS establishes the Master's overriding authority and responsibility to make safety-related decisions. Masters implement the SMS on board; motivate the crew to observe the procedures; issue appropriate instructions and orders; verify that the SMS is appropriate for the ship; periodically review it and report deficiencies to shore-based management. To ensure that the SMS is up to date and appropriate for the ship, you will need to refer to primary sources such as the new Maritime Labour Convention (MLC).

The Master is by law in command and must do whatever is necessary in an emergency to protect and save the vessel, its passengers, crew and cargo. SOLAS Chapter V, Regulation 34.1 states:

Regulation 34-1: Master's discretion

The owner, charterer, the company operating the ship as defined in regulation IX/1, or any other person shall not prevent or restrict the Master of the ship from taking or executing any decision which, in the Master's professional judgment, is necessary for safety of life at sea and protection of the marine environment.

Masters are also obliged under international maritime law to proceed to the assistance of any persons in distress at sea.

Maritime law provides for Masters to give orders to the officers and crew in an emergency that might not be lawful in normal circumstances. Masters can demand that all officers and crew work and obey orders until the emergency is over, in the interests of the safety of the ship, crew, passengers, cargo and others who may be in danger. This right of the Master is maintained in the MLC.

MARPOL

This is the main international convention aimed at preventing pollution of the marine environment by ships from operational or accidental causes.

This convention regulates the discharge of oil, chemicals, harmful substances carried in containers, sewage, garbage and emissions to air. The requirements will be in the ship's SMS manual. Examples of major deficiencies are the absence, serious deterioration or failure of oily-water filtering equipment, inadequate oil discharge monitoring and alarm arrangements and the unavailability or incorrect completion of oil record books.

Masters must ensure that the ship strictly complies with all of the procedures in the SMS for dealing with discharges of oil, chemicals and waste. This convention also recently set limits on the amounts of gases and ozone-depleting substances that may be emitted from ships' exhausts. The ship's SMS may need to be reviewed and updated to reflect these new mandatory requirements.

STCW

The 1978 Convention prescribed minimum standards relating to training, certification and watchkeeping for seafarers. Major revisions to the STCW Convention and Code, known as the Manila Amendments, were adopted in 2010.

The amendments made important changes to each chapter of the Convention and Code, including:

- Revised requirements on hours of work and rest
- New requirements for the prevention of drug and alcohol abuse and updated standards on the medical fitness of seafarers
- New requirements for security training, as well as provisions to ensure that seafarers are properly trained to cope in the event of a pirate attack.

The ship's SMS procedures should have been reviewed to reflect the Manila Amendments to the STCW Convention and your Master's annual review of the SMS should ensure that these have been included.

Maritime Labour Convention

Minimum requirements for providing safe and decent working and living conditions, fair terms of employment, access to medical care, health protection and welfare for seafarers are laid down in the MLC. Masters have overall responsibility for checking that:

- The age, training and qualifications of seafarers meet MLC standards
- Seafarers' medical certificates are valid
- Mandatory onboard complaints procedure is followed
- Ship's safety committee is MLC compliant
- Work schedules safeguard seafarers under 18 and provide sufficient hours of rest for all seafarers and documented records of hours of work and overtime are signed and retained
- Requests for leave and repatriation are handled fairly
- Accommodation and recreational facilities are inspected and properly maintained
- Food is of good quality and quantity and is appropriate for different cultural and religious backgrounds and documented records of inspection are retained

- Provision of health protection, medical care, welfare and social security is appropriate and accessible
- Shore leave is granted fairly.
- As MLC only came into force in 2013, it is expected that PSC inspections will prioritise work and rest hours and individual seafarer records – which were already a requirement of the STCW Convention – and onboard documentation of the formal complaints procedure.

The responsibilities imposed on the Master and the company by the regulations in the conventions are numerous. The SMS must be updated as necessary to reflect any new regulations, codes, guidelines or circulars in coordination with the company, so that the Master can implement them on board.

Managing people on board

By Captain Kuba Szymanski

So, now you are a Captain: congratulations. You have worked hard to get here. It is no accident that you have become a ship's Master. You have been third mate, second mate and finally chief mate. No doubt you were a cadet before that and maybe even an AB.

That means you have had first-hand experience. You have been managed by both the crewing department and previous Masters too. So how was it for you?

In 1999, I was asked: "Are you experienced enough to be a Master? Do you know what to do?" I thought for a moment and realised that probably I knew more about what not to do.

In the shipping industry the top person has to earn that position. Doubtless you have heard many negative stories about it, but we can change that; indeed, you can change that. You have the power to do so.

I was lucky enough to have met great people on my way, people who wanted to share their experiences with me. They were coaching and mentoring me. Do you think you could do the same? Could you share your knowledge? Don't say "It was different then"; that would be a lame excuse. My mentors were real people, just like you.

Have you asked yourself "What type of Captain do I want to be?" If not, maybe you should, if only to help you avoid being 'in between'. This will help your people a lot, because the worst are those who sit on the fence.

So are you going to be an 'iron man' – someone who rules by fear – or are you going to be a 'friend and father'? If you answered 'iron man', then try to remember how you felt when being ruled that way.

Some points to remember:

The people on board the vessel are now your people. You are fully responsible for them. If you look after them – and now as a Master you really can – they will look after you and your ship.

Never ever blame anyone. People are not working with the intention of making mistakes. The fact is, we all make them. Since you are the most experienced person on board you will witness thousands of those mistakes and it will annoy you at times, but you will have great satisfaction when people start making fewer of them.

Always look for the root cause. It is very rare that the true cause of a mistake is an individual. Concentrate instead on systems, environment – surroundings, noise, heat, physical structure – and relationships between people.

Removing someone because they have been involved in an incident, accident or abuse will not stop the occurrence from happening again. That is why you need to remove the root cause.

When the shore office, external or internal audit, port state control or class inspector finds something wrong with the vessel, never pass the buck by saying "This is X's fault, I told him many times". It is even worse if you say that in front of X. No: the buck stops with you. You will have a far more positive impact if you say in front of X, "Yes, this does seem wrong – my fault." You will be surprised how strong that message is. The person responsible for the mistake knows precisely what they have done, or failed to do, and you can be sure they will do everything they can not to make you say that again.

Trust your people. It is very difficult at the beginning, as you probably don't even trust yourself. You have not been Master for long and you worry about everything. It is, however, time to stop micro-managing people. Trust them – without them you will never achieve anything and certainly not the high standards you aspire to.

Invest your knowledge and calmness in your people. You are being watched all the time and your every reaction is being noted. So, first of all, stay calm, avoid snap decisions. That is not a recommendation to take days to reach a conclusion, but just a reminder that only in Hollywood do ships' Masters make instant decisions correctly. We humans need a little longer – and you are one of us.

Respect your people and they will pay you back.

And finally – enjoy, as this is the greatest job in the world!

Managing multicultural crews

By Captain André L Le Goubin

Language barriers are a great obstacle to creating successful multicultural crews, so there needs to be a simple rule: when you do meet, all conversations must be in a common language, usually English. This means that there should be practice in the use of English, so that in emergencies, hopefully the crew is used to talking in that language.

Unfortunately usually little is done to ensure that seafarers can communicate effectively with each other. This is probably the only critical industry where this occurs. Can you imagine an aeroplane where the Captain could not talk to the first officer because they spoke a different language? Inconceivable! Yet I have been on board a vessel recently where the Master could not communicate properly with the chief officer and any instruction, if in any way unusual, had to be written down.

This problem seems to be far more prevalent on vessels with two nationalities rather than those with many, as multinational crewed ships tend to find a common language but, when there are just two different nationalities on a vessel there is a tendency for each nationality to communicate in their mother tongue. Communication between the two different nationalities using the common language appears to happen only when necessary; in essence, depriving the vessel of any social communication between the nationalities. This can lead to all sorts of problems.

The severity of emergencies could be compounded by senior officers communicating in their own languages, so no one else knows what is going on. Only commands are given in English and you can imagine the chaos. This is so unnecessary. If all hands were conversant in the use of English and it was routine for everyone to speak a common language when in the presence of a non-native language speaker then many of the problems associated with a multinational complement would be resolved.

In one vessel with two nationalities the Master only spoke in English when he wanted to include the cadet in conversation. It was like watching a light come on when I started talking about social exclusion due to language and the need for respect for each other. Soon after this very amicable conversation we went down for lunch and I was delighted to see that the conversation was primarily in English, but nowhere near as delighted as the deck cadet who was included in the conversation for the first time, I believe. He just did not stop smiling, engaging and interacting at the table.

A tip – whenever conditions permit, a party or similar social get-together can provide a welcome break from set and stressful routines onboard. Some of the best ones I have been to did not serve alcohol!

Master's standing orders

By Captain Nick Nash

Now that you have the 'top job', remember that you can't be on the bridge all the time. You must rest, but you need your presence to continue to be felt and you need to have confidence in your bridge team to act in your name.

However, if you don't tell your bridge team what you expect from them – they can't act! Writing a set of Master's standing orders is one of the first tasks for any new Master. These days this is probably a set of additional standing orders, as most reputable companies will have already produced their standing instructions incorporated into the SMS.

A Master's additional standing orders need to be brief, to the point and relevant to the operation: they should not make vague references to the 'practice of good seamanship' or 'use all available means to…' We have moved on from traditional phrases and navigation!

All Masters will have a point that they feel comfortable with in traffic and this needs to be made clear to your team. Some Masters may not feel this is necessary as presumably all the bridge officers have a certificate issued by a competent authority, which examines them on the rule of the road. However, if you don't tell the team what you feel comfortable with they will work to their own comfort level which may be less than yours! This includes a bow crossing range and TCPA/CPA, which can be reduced in heavy traffic areas.

I recommend that you should include at least the following in your standing orders, but you should feel free to add your own as you see fit.

Primary navigation

In reality, (D)GPS has become our primary navigation means and radar has become our back up – using parallel indexing and turn monitoring. For this see Captain Paul Chapman's excellent monograph *Monitoring Turns Using Radar* published by The Nautical Institute.

Auto pilot and ECDIS

The standing orders should make reference to the auto pilot/track pilot settings – this is such a vital piece of equipment and the OOWs should therefore have guidance on your required settings.

Continuing this theme, many modern integrated bridges have an ECDIS/radar which can easily be set up incorrectly; this being the case as a new Master you will need to spend some time researching the best settings for your equipment for optimum use and also legal requirements. Some Masters are against being specific in how the equipment is set up, but I feel that the equipment is now so complicated with hidden layers and menus that a standard operating procedure is required. This ensures that each OOW does not inadvertently (or by preference) override your researched safe/optimum settings.

Passage planning

Now that you are the Master, you have the overall responsibility to ensure that the proposed passage plan is safe and can be achieved within your ship's capabilities and, possibly, the charterer's voyage instructions.

The Master and passage planning officer will need to sign the passage plan and add to the voyage portfolio along with:

- ENCs update read me file
- Temporary and Preliminary Notices
- Navigational warnings
- Tidal, current, weather and climate information
- Port information
- Bridge equipment
- Ship characteristics including draught and squat
- Stability plan for ship's stability and loading condition

- Fuel consumption/fuel changeover, usage and availability, emission control areas
- Environmental aspects: discharge, regulatory requirements and sensitive environmental areas such as marine mammal areas
- Security aspects: regulations and security levels for ports, piracy, and terrorist advisories
- Pilot associations
- Port authorities
- Relevant information from ships in the fleet that have performed similar voyages in the past (where available).

Printouts or reference to the following publications:

- Ocean passages for the world (NP 136)
- Routeing charts and/or pilot charts
- Sailing directions or pilot books
- List of lights
- Admiralty list of radio signals
- Mariner's handbook (NP 100)
- Ship's routeing.

This is all good stuff and should be available to the bridge team before and during the passage.

As Master, you will need to run through the passage plan yourself to ensure it is safe, complies with all necessary regulations and is achievable. In particular, I recommend that you pay careful attention to the approach part. Some passage planning officers like to approach the buoyed channel or breakwaters on a tight turn. I prefer the airline approach of a final steady course of 2 or 3 miles, if possible, to get a feeling for any leeway and set before the commit point!

Master's sail plan

I find the best way of checking the route myself is to create an overview of the route, which I call a sail plan. This highlights alter course points, notes key navigational marks to look out for, traffic expected and any particular navigational dangers to be aware of. This satisfies the two points noted above – checking the plan personally and translating a printout into a word picture for the bridge team.

I put my sail plan on a card which I fill in before the voyage and post on the bridge well before sailing so all the team can read and acknowledge it. The sail plan card describes the voyage, but also includes references for planned departure and arrival manoeuvres. A copy of my sail plan can be seen at www.nautinst.org/command

The other side of the card allows for the Master's night orders, in place of the traditional night order book, and allows for any day instructions as well. I feel that use of the sail plan card for the addition of night and day orders is a much more proactive way of passing on instructions to the bridge teams – far outweighing the traditional Master's night order book. Most traditional Masters night orders are brief and actually say little beyond *Comply with all company instructions and call me at…* The sail plan gives very specific details on the actual voyage and allows space for additional detailed instructions over night and day; in fact a live instruction in one document instead of a standalone night instruction book.

However, you must be guided by your own company's SMS in deciding whether to adopt these practices or not.

Navigational briefing

Before the start of the voyage, if your final destination is known, you will need to gather your bridge team and present the voyage as a briefing – a voyage overview – using the ECDIS/chart-pilot display with tracks and depth contours all set, where appropriate. Points covered should as a minimum include:

- Maximum wind and current (speeds and direction) limits discussed and agreed
- Refer to side wind force/tonnes table/graphs (if available)
- Route – speeds – turns – critical points (SBE, FAOP, EOP etc)
- Reporting points
- Whale/fish zones
- VTS coverage
- Traffic including fishing vessel concentrations
- Pilotage
- Ice
- Expected route weather and sources of information
- Major navigational marks
- New wrecks/navigational dangers
- Traffic separation schemes
- Commit and abort points
- Navigational dangers
- Depths/squat
- Environmental lines/zones
- Special areas
- (D) GPS poor signal/loss of signal areas
- Naval exercises/submarine areas
- Working in the chart room.

Good and effective bridge resource management requires the OOW to be front of bridge at all times. However it is recognised that occasionally when in green manning (open sea) one officer may have to do some essential operational work in the chart room. They must first satisfy themselves that it is safe to do so and that any change to the ship's heading/engine setting, configuration, traffic or weather are not expected during any absence from the bridge front.

Remember the safety of the ship has been entrusted to you!

These are my own additional navigational standing orders which take into consideration all the above and may help you design your own.

Master's additional navigational standing orders

MASTER'S COMFORT ZONE

The minimum CPA is 1.0 mile with a bow crossing distance of not less than 1.5 miles and TCPA before taking action of not less than 12 minutes, except that CPAs for vessels of less than 50m this may be reduced to 0.5 miles.

Maximum deviation from track for the purpose of collision avoidance is not more than 2.0 miles and the nearest distance to any hazard to navigation of not less than 1.0 mile.

If it is anticipated that these limits cannot be kept, then you are to call me.

** In the European heavy traffic and routeing/TSS areas the above mentioned CPAs can be reduced to 0.5 mile with a bow crossing of not less than 1.0 mile.

The nearest distance to any hazard to navigation can likewise be reduced to 0.8 mile.

Any changes to this comfort zone will be discussed at the relevant navigational briefing.

TRACK PILOT

As guide for track pilot parameter settings use the NN table, however, the OOW must always use their own discretion in determining the appropriate parameters adapted to the prevailing conditions. Closely monitor settings – switch to heading mode if the system alarms, re-check and adjust settings and ensure ship is steady on course before re-engaging course/track mode.

If alarms continue, engage hand steering immediately and correct any developed ROT with a counter rudder order (consider reducing speed – but not drastically!) Steady ship up and then call me.

Multi-pilot and chart pilot (ENC) settings: My standard operating NACOS set up for each piece of equipment is to be used. It shows the settings that should normally be used for green and red manning levels. Any changes should be agreed by bridge team and noted.

HELM ORDERS

Give concisely and without ambiguity. Prefix a helm order with 'rudder – ROT – steer'

CLOSED LOOP COMMUNICATIONS

Use the interrogative (questioning):
Q: Steer 215?
A: Yes!

Train your helmsmen to do this!

THE PRIMARY NAVIGATION METHOD

Is by (D)GPS on ENCs backed up by real time radar position monitoring techniques (parallel indexing and turn monitoring) and visual confirmation. Any changes to this method will be discussed at the relevant navigational briefing and noted on the passage plan.

We are now electronic navigators backed up by visual clues!

ECDIS TRACK (PASSAGE PLAN)

Printout is to be signed by first officer (passage planning officer) and me before the voyage and to be readily available to the OOW. The turn radius and planned speed for each leg must be shown so as to show the planned ROT which must be within operating limits – particularly in pilotage areas.

BRIDGE ORDER BOOK

I consider the use of my sail plan card to satisfy this requirement for both night and day instructions.

I will file the expired card for reference.

WORKING IN CHART ROOM/SAFETY CENTRE

Bridge resource management requires both OOWs to be front of bridge. However it is recognised that occasionally when in green manning (open sea) one officer may have to do some essential operational work in the chart room/safety centre. They must first be satisfied that it is safe to do so and the front OOW must call the back OOW to the front of bridge before any change to the ship's heading, engine setting, configuration, traffic or weather. Remember the safety of the ship has been entrusted to you!

REMEMBER

Courtesy and respect to other departments. Be polite on telephone and radios and keep the ECR and others advised of schedule and weather changes.

Nash, Master

A very senior Master once said to me that being a Master was a 'mortal responsibility' and that there is no other job in the world like it. He was right but you still have to try to put the job into perspective. I do believe we need to move forward and realise that our electronic navigational systems are no longer aids to navigation, but the main method of navigating, with most importantly, visual confirmation. The above standing orders, standard set up of equipment and finally the sail plan recognise this. I have tried to move away from vague generic terms and stick to the point. Your bridge team needs specifics and need to know what you feel comfortable with (your comfort zone) so they can act in your name and call you in good time with accurate information.

A good bridge is one that is relaxed but professional and operates as a team to standard operating procedures and instructions.

Relationships with stakeholders

Who are the stakeholders?

By Graham Cowling

The Master's relations with the various stakeholders involved with the ship's voyage are complex. You might be in command of a vessel owned by a holding company in Liberia, flying the Liberian flag. The owner might be an investment fund in New York that has contracted a third-party ship manager in Hamburg to do the technical and operations management. This manager is the Document of Compliance (DoC) holder and has placed the crew supply and management contract with a crewing agency based in Cyprus. The vessel is classed by a Japanese classification society. The charterers are located in Tokyo. Hull and machinery cover is provided by insurers based in Oslo and P&I cover by an insurer headquartered in Bermuda. The vessel trades between the United States and China. The crew is multinational, drawn from six countries.

Each stakeholder expects the Master to look after the vessel, sailing it without any breakdowns, delays, pollution or accidents, and to follow a tightly controlled budget. Other stakeholders include ports, pilots, terminals, repair companies, suppliers and salvors.

Stakeholders' expectations of the Master may sometimes be at odds with each other. If so, you should be very careful.

Respond quickly to communications received. If you need time to collect information, reply immediately, promising to get back later with details. A quick response conveys a good professional image and the impression that you care. Most stakeholders are accustomed to calls in the middle of the night, so if the situation demands it, make that call!

Communication should be clear and relevant, keeping in mind that the other person needs to understand the issue fully. Include photographs and drawings where necessary.

Do not be offended when asked questions. Questioners have their own responsibilities and may not have marine experience. It is their job to ask and the Master's to respond.

Use the support systems available. Stakeholders often complain that Masters do not ask enough. Pilots are uncomfortable when a Master does not ask questions about the manoeuvre. DPAs are unhappy when a particular action is taken by the vessel without consulting the office. Stakeholders prefer to be called by the Master about a developing situation rather than receive a nasty surprise by email. Always use professional but assertive language and show understanding and respect for diverse cultures and personalities.

The DoC holder will expect the Master to operate the ship safely and in full compliance with the ISM Code and other laws and regulations intended to prevent loss of life, damage or pollution. Very close cooperation is essential between the ship and the DPA on all safety-related issues and any concerns the Master might have. If there are conflicts of interest between stakeholders, you will need to discuss them with the DoC holder.

The vessel's owner may be an investment fund or a bank, or a traditional shipowner who has contracted technical management to a ship manager, meaning the Master may not be in regular touch with the owner. To protect the owner's position as much as possible, the Master must make every effort to avoid off-hire, unscheduled breakdowns and delays, as well as ensuring that the vessel is properly managed and maintained.

If, for example, the charterer requests the vessel to reach the next port at a specified time, this might require steaming at speeds that are not safe in current weather conditions, creating a risk of damage to the ship and cargo or injury to the crew.

The charterer expects the Master to operate the ship without any delays or off-hire and with maximum flexibility for cargo intake and schedules. Prompt and accurate reporting is expected. The Master must comply with such requirements while keeping in mind that safety of the vessel, crew and cargo takes precedence over all commercial considerations. Do not hesitate to contact your DPA if worried that a requirement might jeopardise safety.

Other stakeholders will impose safety- or environment-related requirements on the vessel. Port state control, coastguard or environmental inspectors will check compliance when on board, but Masters have a responsibility to the owner and manager to ensure those requirements are correct and applied within the regulations.

Hull and machinery (H&M) and protection and indemnity (P&I) insurers will expect the Master to take all steps to minimise loss in the event of a casualty and to act as if the vessel were uninsured.

The Master should expect support and clear information and advice from stakeholders. The DPA should be available at any time day or night on the telephone or by email. The Master should be in no doubt about whom to call in such cases. Technical managers can provide support and advice on the running of the ship.

Section 1
The Nautical Institute on Command

The Master should receive clear instructions from the charterer's operations department on matters such as voyage orders, bills of lading, details of cargo, hazardous cargo and special loading instructions.

The Master is in command and is given an overriding authority for the safety of the ship and crew under the ISM Code. Whatever the orders or advice given by stakeholders, the Master retains this authority, which should not be used lightly.

You are in charge whether the vessel is at sea, alongside, berthing or at anchor, and you need to be assertive. Always convey your views to the stakeholder, however uncomfortable it may be for the listener, as your actions or inaction will determine whether the vessel operates properly. For example, even with a pilot onboard the Master remains responsible for the safe navigation of the vessel (except in the Panama Canal). If something concerns you, speak up and take over if necessary.

At the start of each charter, read the key parts of the charterparty (CP) or recap and make sure you understand them. For example, on a container ship, who pays for lost lashing gear and how? Is it compensated on a lump sum or actual cost basis? Who pays for sludge removal? How much is the communication/entertainment/victualling (CEV) allowance? More importantly, is the ship exceeding the allowance given by the charterer? While some owners may not provide all the CP information terms to the Master, you should ask for the sections of the CP that might affect the correct operation of the vessel.

The owner makes a warranty (similar to a guarantee) in the charterparty that its vessel will perform at a certain speed and consumption. If this warranty is not met then the owner must compensate the charterer for the time delay (lack of speed) and excess bunkers (over-consumption). This warranty only applies weather and safe navigation permitting (WSNP). Discuss this with the technical managers so that you are completely familiar with the requirements.

In the case of a cargo damage incident on the vessel, both charterer and owner will probably send their P&I club surveyors to conduct a damage survey. To protect the interests of all parties, the Master should be very careful about the type of information disclosed to each surveyor. You need to know exactly who will be involved and should obtain clear instructions from the technical managers and charterers about who should be allowed on board and what they are permitted to survey.

The type of information passed to the agent will depend on whether the agent is the charterer's agent or the owner's husbandry agent.

A review of the pre-load manifest might show that the vessel is going to load overlength or oversize containers, containers full of calcium hypochlorite, iron ore fines, direct reduced iron (DRI) or a cargo of steel coils. All of these cargoes present special hazards for an unprepared Master. Damages or losses affecting cargo interests may result in huge claims to the owner. You should always consult the technical managers or charterers if unsure. Pre-load surveys will probably be needed in the case of steel cargoes and hazardous bulk cargoes such as iron ore fines. Special declarations may be needed to confirm the cargo is safe to load.

Disputes often arise between owners and charterers about how damage occurred. Successful resolution for the owner frequently depends on the Master's ability to prove that the damage arose from the negligence of crane drivers or stevedores. Evidence needs to be gathered quickly and secured so it can be used to settle a dispute. For further guidance see The Nautical Institute publication *The Mariner's Role in Collecting Evidence in the Light of ISM*, third edition (2006) and *Handbook* (2010).

Relationships with flag state

By James Parkhouse

A primary responsibility for Masters in their relationship with the ship's flag state is to ensure that the vessel is maintained to meet statutory requirements. Here we look at where those statutory requirements come from, how they are applied and what you as Master should do if you find that your ship is not, or may not be, in compliance.

Each flag administration has its own policy on how to implement IMO requirements and guidelines and on ensuring acceptable levels of safety and compliance. Each administration should have set out the practical implementation of its national legislation and policies in guidance documents, which are often supported by a concise handbook or guide to national legislation for day to day use by ships' Masters. It is important that Masters have access to, and are familiar with, these sources of information.

The ISM Code requires that the company's safety management system (SMS) takes account of flag administration requirements. As part of the Master's periodic onboard review of the SMS, you should report any shortcomings in this respect to the company so that corrective action can be taken.

A flag administration has various ways by which it can ensure its ships comply with statutory requirements. First, it ensures that its ships are surveyed and certified in accordance with statutory conventions and national regulations. Such surveys may be conducted by the flag administration or, more usually, by a recognised organisation (RO). The RO will generally be the ship's classification society, which is authorised to act on behalf of the flag administration to conduct surveys and issue certificates. Second, the administration carries out its own annual safety inspection, which is separate from the surveys carried out by the RO. It gives the flag a snapshot not only of the condition of the ship but also the quality of the work being conducted by its ROs.

It is not possible for all flag states to monitor the ships under their flag at all times around the world and PSC is a means of regulation that should be familiar to all Masters. Although often maligned and seen negatively by many in the industry, PSC is, in effect, a mechanism by which other administrations may reassure the flag state that its ships are operating in compliance with the appropriate codes and regulations. When applied in a fair and equitable manner, PSC should be seen in a positive light as helping Masters with the effective and safe management of their ships. The scope and extent of PSC inspections varies with ship type, age and period since last inspection, and the region in which the ship is operating. The initial impression the ship makes on the PSC officer may also influence how deeply the officer decides to inspect the ship.

Flag administrations and classification societies publish guidance to assist Masters in the effective management of PSC inspections.

It has to be recognised that equipment breaks down and accidents happen from time to time.

When deficiencies are identified at surveys, safety inspections or PSC inspections they may be dealt with in a number of ways. For minor matters, the flag administration may grant a short period in which the deficiency can be rectified. The affected statutory certificate may be issued with a limited validity provided temporary measures are put in place. For more serious deficiencies, the ship's departure may be delayed until the matters have been rectified. At a PSC inspection, a PSC detention may be imposed to prevent departure of the ship.

In addition to delaying the departure of the ship a PSC detention can damage its future commercial prospects. Even in instances where a detainable deficiency is identified and rectified in the space of a few minutes, perhaps while the PSC officer is still on board, the ship will nevertheless be detained, released and the detention reported to the regional PSC body and the IMO. Detention records are publicly available, so can be checked by charterers looking to fix vessels. A ship that appears to present a risk to the shipper or charterer is unlikely to attract the best cargo or hire rates. Time off hire could lead to loss of the charter or a claim by the charterer against the owner. The ship may also be targeted by PSC for more frequent and thorough inspections in future.

The potential long-term loss to the shipowner, both in reputation and income, can far outweigh the cost of resolving the deficiency or ensuring it does not occur in the first place.

Given that Masters have a responsibility to ensure that ships are correctly maintained in compliance with statutory requirements, and bearing in mind the consequences that can arise from non-compliance, Masters clearly need to do everything possible to prevent deficiencies arising. If the Master and owners report defects or deficiencies to the flag administration and to the ship's RO at the earliest opportunity, then the flag administration and RO can put in place appropriate measures, either permanent or temporary, which should allow the validity of the statutory certification to be maintained. Obtaining such assistance from the flag administration and RO may well prevent intervention, and possible detention, by PSC.

Relationships with class

By Peter Hamer

Classification societies have a variety of roles in the maritime industry. They provide classification and can be authorised by flag states or other governmental organisations to carry out statutory services. They also have a role as recognised organisations (RO) under international conventions such as SOLAS. These permit flag states to delegate the inspection and survey of ships to RO which must follow the specific regulations and guidance for that flag.

Classification societies also assist the maritime industry and regulatory bodies on matters relating to maritime safety and pollution prevention. They are often simply referred to as class societies or just class.

The objective of ship classification is to verify the structural strength and integrity of essential parts of the ship's hull and its appendages, the reliability and function of the propulsion and steering systems, power generation and those other features and auxiliary systems that maintain essential services onboard. Class societies achieve this by codifying and maintaining their own rules and verifying compliance with them through surveys documented by a classification certificate. The rules are developed over many years through extensive research and the collection of service experience and are constantly refined.

However, a classification certificate is not a warranty of the ship's safety, fitness for purpose or seaworthiness. It is an attestation that the ship is in compliance with the applicable rules. The classification of a vessel is based on the understanding and rule requirement that the vessel is loaded, operated and maintained properly by competent and qualified personnel.

The classification process consists of:

- A technical review of the design plans and related documents for a new vessel to verify compliance with the applicable rules

Attendance by class surveyors at:

- The shipyard during construction of the vessel to verify that the vessel is constructed in accordance with the approved design plans and classification rules
- The production facilities providing key components such as the steel, engine and generators, to verify conformity to the applicable rules
- The sea trials and other trials of the vessel and its equipment before delivery, to verify conformity to the applicable rules.

When all of these have been completed successfully, the assignment of class may be approved and a certificate of classification issued.

Once in service, the owner must submit the vessel to a programme of periodical class surveys to verify that the ship continues to meet the rule requirements for continuation of class. Overdue surveys can result in automatic class suspension for a vessel.

When a ship is suspended or withdrawn from class, the society notifies the ship's flag administration and, as a consequence, the flag administration usually invalidates the statutory certificates, effectively stopping the ship from trading.

SOLAS and other international conventions cover many areas (see pages 5-8). Some or all of these may also be covered in a particular class society's rules.

It should be emphasised that the safety and integrity of a vessel is the responsibility of the shipowner, and, on a day to day basis, the Master. The effectiveness of the classification system for vessels in service depends upon the shipowner cooperating with the class society in an open and transparent manner on all issues that may affect a vessel's class status.

Should defects, damage or deterioration that may affect class become apparent between surveys, the owner or its representative must inform the society without delay.

Copies of the basic class rules governing this responsibility are readily available via the internet or from visiting surveyors. These online rules will improve a Master's understanding of the requirements and criteria used when establishing non-compliance and also of the survey systematics.

A classification surveyor may only go on board a vessel once in a 12-month period. The surveyor is not expected to scrutinise the entire structure of the vessel or its machinery. The survey involves sampling, for which guidelines exist based upon computation, empirical experience and the age of the vessel. The guidelines indicate those parts of the vessel or its machinery that may be subject to corrosion, are exposed to the highest incidence of stress, or are likely to exhibit signs of fatigue or damage.

Masters should have a good knowledge of these surveys, their due dates and validities. Before a survey takes place Masters should familiarise themselves with the survey's scope and any mandatory pre-survey preparations such as tank cleaning. All these requirements should be discussed and clarified in a pre-survey meeting.

Planning by the ship's management team and its cooperation are crucial to the effectiveness of a survey. Survey planning should ensure that adequate preparation is made and that the conditions and timing allow for the work to be carried out effectively. Plans might consider, for example, whether water visibility, weather and tidal conditions are adequate for an underwater survey.

It is essential that classification surveyors are made aware of crew working hours, cargo activities, unsafe spaces or activities. Particular care should be taken when planning surveys that require access into enclosed spaces or access at heights.

After the survey, the Master and surveyor should have a common understanding of any survey findings, which should be properly documented. This is essential so both shipowner and class society are clear about what is required and expected. Special attention should be paid to the nature of the findings, as different class societies use different terminology for items that must be rectified as opposed to those that have been recorded as observations.

The Master should double-check that all certificates are issued or endorsed correctly with signatures, stamps and dates along with accurate details of the vessel. Small errors can lead to large consequences. For example, an incorrect address of an owner on an ISM safety management certificate (SMC) can lead to a detention.

It should be remembered that the classification surveyor's primary goal is to ensure the safety of the vessel and crew – an objective that is very much aligned with the Master's.

Relationships with charterers and owners

By Captain Ghulam Hussain

A vessel performing under a charterparty is subject to a host of obligations, responsibilities and warranties. The Master's role, while primarily to protect the owner's interests, also includes looking after the charterer's directives and other interests from third parties such as shippers and receivers who may not necessarily be the charterer themselves.

This multifaceted role needs to be performed professionally and diplomatically. Much depends on a Master's ability to deal conclusively with the varied situations that arise during the charter. As Master, your role is pivotal, and often a timely act can resolve a dispute between owner and charterer.

The charterer usually chooses a suitable charterparty for the type of vessel and nature of the trade. Common dry cargo charterparties are the Bimco GENCON 94 for voyage charters and NYPE 93 for timecharters. It is not uncommon to see charterparties with some 100 rider clauses in addition to the main terms printed in part 1 of the document. These act as clarifications to the main body of the charterparty and form an integral part of the charterparty itself. The Master must ensure timely compliance with all these riders.

Despite the riders, which are drafted and agreed with the aim of avoiding subsequent disputes, there remains a risk that charterer and owner will interpret the agreement in different ways.

Even the most efficient of Masters can find themselves dragged into disputes and controversies that arise between the owner and charterer. A breach of the terms, conditions and exceptions of the charterparty to the extent that a compromise cannot be negotiated may necessitate long arbitration proceedings.

This article does not deal with disputes or the intricacies of the various types of charterparty, but simply provides a checklist for the Master's guidance at each stage of the charter.

Before vessel's delivery to charterer

- Ensure a copy of the charterparty – or at least the main terms or fixture note – has been received from the office

- Confirm that the voyage instructions have been received from the charterer and formally acknowledge their receipt. In the event of any conflict between the charterer's voyage instructions and the charterparty terms, seek clarification from the office
- Keep handy the contact details of all parties involved in the voyage, including charterer, port agents and P&I correspondents in the expected ports of call
- Request weather routeing company details from either the owner or charterer
- Send passage plan showing waypoints, distances, steaming time etc. to the charterer, even if voyage instructions do not require this
- Send the charterer the cargo loading plan, stowage plans and departure and arrival draughts in salt and fresh water, with details of density, freshwater corrections and consideration of load line zones.

Delivery

- Ensure the notice of readiness (NOR) is tendered correctly in writing. Check whether the vessel is an arrived ship and whether NOR can be tendered at any time or during a specific period. Many disputes arise from flawed NORs
- Ensure all parties are informed in writing, particularly charterers and shippers (normally via nominated agents)
- Prepare for on-hire condition survey (by charterer initially), hold inspection and bunkers on delivery survey. Other surveys may be ordered by shipper, receivers or port state control (PSC) surveyors. Ensure that bunker levels on delivery are accurate when delivering on a timecharter.

Loading port

- Send daily report of loading to charterer
- Ensure correct and timely entries in log books
- Prepare for surveys, such as ISM, PSC and class
- Prepare a correct statement of facts. If this is prepared by agents, request a copy for scrutiny well before vessel's departure
- Deal politely but firmly with Port Captains or other charterer's supervisors
- Always check draughts, trim and stability
- Issue letters of protest where necessary, especially regarding stevedores, stowage, dumping of cargo, damaged cargo, misuse of cargo gear or any breach of charter terms, such as the vessel not remaining afloat at berth when this was specifically agreed
- Sign mate's receipt with care, adding clauses if necessary
- Check daily stevedore reports, adding remarks where necessary
- Authorise agents to sign bills of lading on behalf of the Master only after office approval is received
- Appoint P&I surveyor if required while loading. For steel cargo, this would be for a pre-shipment inspection; for bulk grains, to attend to draught surveys before starting and after completing loading; for bagged cargo, additional tally

- Contact P&I club and owner if discrepancies between draught survey quantity and shore terminal quantity are found, or for a dispute over bunker figures (especially after bunkering on charterer's account)
- Refuse any letter of indemnity from any party in the absence of the owner's or P&I club's approval
- Ensure that the ship and crew abide by ISPS regulations and local port security rules.

The voyage

- Noon reports to be accurate and in charterer's required format. Additional reports in the company format can be sent to office separately. It is good practice to copy the owner in on messages sent to the charterer
- Provide a correct declaration of daily fuel consumption (including BROB) and speed averages
- Send accurate declarations of weather parameters to the nominated weather routeing companies. Do not exaggerate the weather experienced nor conceal facts like stoppages, however brief, as this can lead to disputes over potential off-hire periods or vessel non-performance
- Ensure there is a good match between engine room log book and deck log book entries on timings, bunker figures, engine movements, slip and speed
- When declaring bunker consumption or BROB, avoid the practice of keeping fuel in hand for adjustment later
- Regularly sound tanks and conduct fire and safety patrols
- Attend to cargo ventilation requirements if necessary
- Send accurate ETAs to concerned parties
- Send correct and timely declaration of all pre-entry and inward documents to agents
- Ensure vessel arrives at even keel, correcting the trim maintained at sea
- Ensure ballast control is maintained on departure and arrival and throughout the voyage.

Arrival at discharge port

- Tender accurate NOR on arrival
- Attend carefully to draught surveys on arrival. There may be a discrepancy in the cargo quantity. It is advisable to have the owner's P&I surveyor in attendance
- Attend to off-hire bunker survey. The charterparty may call for a joint survey and the charterer will nominate a surveyor for owner's acceptance. Ship's staff should remain vigilant and witness the off-hire bunker soundings for accuracy
- Before permitting discharge, ensure original bills of lading are presented by the receiver or the receiver's agent and have been checked before opening hatches
- Do not discharge cargo against a letter of indemnity, unless approved and instructed by the office
- Deal with potential cargo damage and short landing claims. Keep photographic evidence on file
- Give a redelivery notice or completion of voyage notice to the charterer.

Some points to remember

- Use the Julian calendar when calculating the days between dates in different months
- When entering timings, use tenths of an hour, ie 6-minute intervals. This makes it easier to calculate total steaming time and allows calculations to the nearest decimal point. For example, writing 15.00Z or 15.06Z as full away on passage is better than 15.01Z. NB: this rule should never be applied when entering timings of an accident, collision, position fixing etc. It is good practice to state times in both local time and GMT, ie 16.00 LT (GMT -4)
- Calibrate hydrometers and keep certificates up to date. During a draught survey, three surveyors having differing hydrometer readings could delay a vessel's departure by hours haggling over a cargo quantity difference of 50 tonnes
- Beware of certificates given by shippers showing transportable moisture limit (TML), especially when dealing with cargoes like nickel ore
- Anticipate a potential dispute beforehand. If, when loading from shore terminals, three consecutive draught surveys show quantities differing from shore figures, contact the owner rather than wait for a dispute to arise. Weighbridges may also give differing quantities. In these circumstances, the charterparty terms are critical.

And finally, when you are exhausted and thinking of resting after a river pilotage of 36 hours, all shore requirements completed, initial draught survey completed and cargo loading started, you may get a call from the office asking when the pilot disembarked, or from the charterer asking what would be maximum loadable quantity for departure 08.00 next day on a falling tide. Don't worry; it's all in a day's work!

Relationships with pilots

By Captain Sean Bolt

Much has been written and debated about the nature of the relationship between the Master and the pilot. When underway with a pilot on board you as the Master have command and the pilot has the con of the vessel. The Master retains overall responsibility for the vessel and, in some jurisdictions, will be held fully responsible in the event of an accident – even if caused by the fault of the pilot, unless the pilot is proved to have been negligent.

You inherit a relationship established by historical precedent and practicality. Like any relationship it can be good, bad or indifferent. The actions of both the Master and the pilot can go far to ensure a positive, safe and efficient evolution.

Ports insist on compulsory pilotage because they have expensive and critical infrastructure to protect. In most cases the pilot will be a more experienced shiphandler than the Master. The pilot understands local conditions and will do the job efficiently. So how does the Master ensure a positive relationship?

Section 1
Relationships with stakeholders

Your first step is to ensure your vessel is ready to receive the pilot at the appointed pilot boarding area. Arrive on time, at the speed requested and with a properly rigged pilot ladder, to the latest IMO standards, at the height above the waterline requested. If you are at an anchorage and have been requested to be underway, don't still have five shackles in the water when the pilot is expected to board. If you have been asked to remain at an anchorage with two shackles on deck, don't think you are doing the pilot a favour by being underway. Be prepared to create an adequate lee for the pilot without putting your vessel at risk. Bridge windows should be clean and compass repeaters and instruments on bridge wings should be uncovered and readable. Welcome pilots to the bridge and offer some form of refreshment – they may have been on their feet for a long time without sustenance.

It is likely that the pilot does not speak the same language as you. If using English as the common language take care that all orders given by the pilot and any responses by you are fully understood by all on the bridge.

Clearly point out to the pilot all pertinent bridge equipment – compass, radars, rudder indicators, speed logs, echosounder etc.

Know your draught and your compass error. If something is not working or the compass has an unacceptably large error, tell the pilot. Pilots will be less concerned with the rights and wrongs of what should or should not be working than with knowing what they can rely on during the pilotage. Know the constraints on engine manoeuvrability and where the critical revs lie. Ensure the person on the wheel is experienced and will understand the helm orders given by the pilot. You should expect from the pilot a comprehensive passage plan. Understand that plan so you can monitor progress.

When going over passage plans pilots should always remind Masters to speak up if they feel uncomfortable with any aspect of the passage.

The Master is also expected to be aware of the latest bridge resource management techniques. The Master should understand the power of tugs – how, when and where they will be used. This is something the pilot should explain.

Be sure to explain to the officers the intended tie-up and the order of tie-up. If a mooring boat is being used, understand and explain to the deck officer which lines to run first and how many can be put into the mooring boat at a time. If ship's heaving lines are to be used, ensure the crew has them ready and available. Don't give the instruction to send lines until told by the pilot and also instruct crew members not to heave up mooring lines until instructed by the pilot.

English is the international language. Any conversations by the pilot over the radio in a different language should be explained and clarified by the pilot, especially if it involves the passage plan and any alterations.

If you feel uncomfortable at any point during a piloting passage, tell the pilot directly about that uneasiness. This should result in the situation being put right immediately. If not, take it up a notch and clearly tell the pilot that you are not comfortable and that something must be done immediately. Try to avoid Master/pilot confrontation, as this rarely ends well for either party.

However, if you are going to override the pilot, then say so clearly and explain why. For the pilot, there is nothing worse than finding that an engine or helm order has not been carried out because the Master has overridden this order (perhaps in another language so the only indication to the pilot is sight of the engine revs or rudder indicator), especially when that order is critical to the efficiency of a particular manoeuvre. Either you or another officer on the bridge should monitor the pilot's helm instructions and ensure that the helmsman carries out those instructions. Use closed-loop communications.

Don't leave the bridge once the vessel is in a channel or turning basin or being manoeuvred alongside. You may have reporting requirements to the owner and charterer, but those requirements will never outweigh your responsibility as Master to remain in command of the vessel. Being in command means you need to be on the bridge monitoring what is happening.

What should the Master expect from the pilot? A courteous and professional attitude, followed up by a safe piloting evolution, in control at all times and as per the passage plan. The pilot should also understand that the Master/pilot exchange is continuous, because important information should be exchanged throughout the passage. Orders from the pilot should be clearly audible, concise and understandable to everyone on the bridge.

By building a positive relationship with the pilot and working together, you will enhance the safety of your vessel and improve public trust.

Relationships with port state control

By Captain Neil Forde

This is not a detailed study of port state control as practised across the globe. What it strives to be is a brief introduction to its mechanism and consequences with some advice on how best ships' Masters should approach it.

The fundamental purpose of port state control is to ensure that ships calling into ports comply with international conventions. SOLAS Regulation 19 sets out the guidelines and reasons for port state control. Similar instructions to contracting parties are found in MARPOL Chapter 2 Regulation 11.

Some states may also check for compliance with national legislation during a PSC inspection. Your ship's agent will inform you of specific local regulations.

Port state control organisations have developed on a regional basis, usually under agreements called memorandums of understanding (MoU). For instance, the European Union, Northern Europe, Canada and the Russian Federation are covered by the Paris Memorandum of Understanding (PMOU).

The trend is for more cooperation between the MoU organisations which exchange information globally through the Equasis database. The Paris MOU and Tokyo MOU (Australasia and States of the Western Pacific region) also carry out joint concentrated inspection campaigns.

The best preparation for a port state control inspection is to always be ready for scrutiny as these may be time triggered but may also be caused by an event which the Master may or may not be aware of. These are referred to in the PMOU as unexpected factors or overriding priority events. A close quarter's situation report, an observation by vessel traffic management systems, a report from a pilot or stevedore or apparent erratic behaviour by the vessel are some examples of events that may bring about a PSC inspection.

On first taking over a vessel carry out your own inspections following the PSC format. Check the vessel's statutory certificates are all immediately available and valid. If a certificate appears to be missing or invalid check with the departing Master that it has not been misfiled and, if not available, immediately contact the shore management and notify them of the situation.

Check all the crew certificates and ensure that all short course certificates are valid and available in the original format. Compare the STCW watchkeeping certificates with the minimum safe manning requirements for the vessel. Check all medical fitness certificates and ensure that any crew members approaching expiry of a medical fitness certificate are scheduled for a medical examination.

It will only take about 40 minutes for port state control officers (PSCO) to carry out these checks on most vessels, which leaves them plenty of time to inspect other areas of the ship.

Your second step is to gather the ship's management team and walk about around the ship bringing fire-fighting and lifesaving appliance training manuals with you. Look for defects and check that the equipment onboard matches the descriptions in the manuals. Checklists are provided by many class societies and P&I clubs. Include senior officers so that they, in turn, motivate junior officers and other crew. The engine room and auxiliary engine rooms must be part of this inspection as Masters are responsible for their condition.

These inspections should ideally be conducted monthly and always at the beginning of your appointment and before leaving the vessel. On a hard run vessel it may be hard to find time for such a full inspection so break it down into manageable chunks. Compare your inspection findings with the vessel's own non-conformity log and get an explanation for outstanding items awaiting rectification from your own staff and from the shore management.

So you think you are ready for a port state control inspection?

Once a PSCO is onboard, alert all personnel that a PSC inspection has started. There is no need to surreptitious about this. Declare to the PSCO any work currently being undertaken that may disrupt the normal operation of the vessel: life rafts being replaced; overhaul of a major element in the engine room; hot work in a cargo hold etc.

Present your ship's certificates in an orderly manner. Make sure invalid and expired certificates have been removed from the certificate file. Poor document control is considered a deficiency. Inspectors will have formed opinions of your ship by the time they have arrived in your office and the presentation of certificates in an efficient manner will demonstrate your familiarity with the documents.

All personnel in a position of responsibility should be fully familiar with the ship's safety equipment and be able to demonstrate its use. Many Masters insist on accompanying the PSCO and this is good practice; however it is very impressive when a junior officer undertakes this task and can demonstrate a thorough understanding of the vessel and its operational systems. It is not so impressive when the Master is unable to demonstrate such ability and has to ask a junior officer to explain, for instance, the operation of the CO_2 system. This could alarm the PSCO who may escalate the inspection.

Some bad habits and practices are so familiar and accepted on some vessels that crews do not take the simplest steps to remove these faults. Masters and management teams should be alert for these habits and practices on your inspection and ensure that if found the situation is corrected and the crew made aware of the reasons for these features in the vessel design.

PSCO report and list of deficiencies

Make absolutely certain that you fully understand the deficiency being given and why. Ask for the convention reference (this is now required on all PMOU reports of deficiencies), particularly if you feel the deficiency is not deserved. This can then be used as basis for an appeal. Do not get into a confrontation with the PSCO. They should behave in a considerate and professional manner and expect to be treated similarly by the Master and crew of a vessel they are inspecting. If a PSCO does behave badly or erratically, contact your shore management and pass on their office details.

PSCOs do make mistakes. Usually this is in the interpretation of the regulations. If the matter cannot be settled onboard then it will be necessary to refer to the owner who should request that the flag state intervene.

When a vessel is detained it is vital that the deficiency is fully understood and the Master and crew should make every effort to deal with it as quickly as possible. Masters should take careful notes of discussions with the PSCO and details of the deficiency should be clearly stated in writing. Photographs should be taken and kept in the original format. All relevant information should be catalogued and kept safely as it may be required in the event of an appeal by the owner.

Above all stay positive during the PSC inspection. Detentions are rare and very unusual on a well-managed ship. Many ships record no deficiencies after repeated inspections. In the vast majority of cases the PSCO is concerned for the safety of you and your vessel and if a good rapport is achieved you will learn more than you expect.

Relationships with insurers

By Chris Adams

Merchant shipping has always been, and remains, an enterprise beset by risk, notwithstanding advances in ship design and construction, technology and training. That continuing risk, together with the ever-increasing values of vessels, their cargoes and bunkers, emphasises the need for sound insurance. In its absence, the potential exposure arising from damage, loss or liability could threaten the company's existence. Consequently, every well-found shipping company has in place a comprehensive package of insurances to protect against the full range of risks it may face.

The Master needs to be aware of insurances concerning the vessel itself and the liabilities that might arise from the vessel's operations – respectively, the hull and machinery (H&M) policy and protection and indemnity (P&I) cover. These policies insure against the consequences of marine perils such as heavy weather, fire and collision. War and terrorism risks are excluded from H&M and P&I policies, so separate cover is required. If the vessel is trading in an area where there is an increased risk of piracy, the owner may have arranged kidnap and ransom (K&R) insurance as an adjunct to the war risk policy. Since there is often an obligation not to disclose the existence of such cover, Masters are unlikely to know if their vessel and its personnel are covered by K&R insurance.

Standard forms of policy wording generally apply in respect of H&M insurance and depend very much upon the market in which the shipowner has chosen to place the insurance. For example, in the London market, the Institute Time Clauses (Hulls) may apply, but elsewhere other forms of policy wording may be used such as the Norwegian Marine Insurance Plan or the DTV-German Standard Conditions of Insurance. Masters should familiarise themselves with at least the outline framework of various forms of H&M insurance to gain an understanding of the risks covered.

For P&I risks, most of the world's merchant fleet is covered by one of the 13 members of the International Group of P&I clubs. The clubs originated as regional associations of shipowners that insured, on a mutual basis, categories of liability risk that had either been excluded from H&M policies or had emerged from legislative development. International Group clubs still provide P&I insurance on a mutual basis and are governed by boards of directors drawn from their shipowner members. Some commercial insurance companies provide P&I cover on a fixed premium rather than mutual basis. They offer lower limits of financial cover than the mutual system provides, so they mostly attract vessels of smaller gross tonnage.

While the general focus of the H&M policy is upon the vessel itself and its machinery, it often also includes some coverage in respect of collision with other vessels. Under the Institute Time Clauses, for example, three-fourths of the collision liability is covered by H&M underwriters. Historically, it was this limit of three-fourths collision liability under H&M cover, and the consequent uninsured exposure in relation to the remaining one-fourth, that prompted formation of the P&I clubs. Under this traditional arrangement the collision risk is shared between two policies, but the H&M policy limits the extent of collision cover to the insured value of the hull. In cases where the three-fourths exposure proves greater than the insured value, the P&I cover takes care of that excess. There are also certain categories of liability arising from a collision, such as oil pollution or wreck removal that are expressly excluded from the H&M cover, and again the P&I cover will respond to these.

Sometimes the H&M policy also covers certain types of damage to fixed or floating objects. Alternatively, the collision liability might be excluded completely from the H&M policy and covered in full by the P&I club. The intention is always that the terms of the H&M and P&I covers should dovetail together to avoid any gaps in insurance protection. The vessel will generally carry onboard a copy of its certificate of entry with its P&I club and this should reveal the extent to which collision liability is covered, or any exclusion from the normal extent of cover that might apply.

Salvage is another area where both the H&M and P&I insurers are likely to be involved. If salvors are required to preserve the vessel, cargo or other property and her crew from loss, a successful salvage operation conducted under a Lloyd's Open Form (LOF) of salvage agreement will result in an award in favour of the salvors. The amount of that award will be determined according to the value of the property salved, the dangers faced and the degree of skill exercised by the salvors in conducting that operation.

Masters should remember that they are ultimately responsible for their vessel, its cargo and crew and should have no qualms about engaging salvage assistance on LOF terms if necessary, even if there is no opportunity to consult the shore management team beforehand.

Where salvage is successful, the salvage award will be payable by the owners of the property salved – typically ship, cargo and bunkers – and covered by the insurance policies on that property. The cargo owners and their underwriters might seek to recover from the owner what they have had to pay in respect of salvage should they be able to establish a breach of the terms of the contract of carriage. Typically they will aim to prove that the owner failed to exercise due diligence to make the vessel seaworthy at the start of the voyage. If such a breach is established and the owner is liable to pay the cargo owners' proportion of salvage, that liability will be covered by the P&I club. In view of this, the club is likely to investigate any incident involving salvage services so as to determine its potential exposure.

In cases where the ship cannot be salved and becomes an actual or constructive total loss, a liability to remove the wreck or to clean up oil pollution may then arise. The P&I club will cover such liabilities.

The scope of the liabilities covered by a vessel's P&I insurance is wide, but most claims involve loss of or damage to cargo, crew illness or injury, personal injury to third parties, collision, damage to fixed or floating objects, and pollution and wreck removal.

Where an incident occurs that may potentially involve any of the vessel's insurances it is vitally important to notify the owner promptly. The owner will then contact the relevant insurer so that it can set in motion an appropriate response. This usually involves the appointment of a surveyor, who will attend the ship to inspect either the damage to the ship or that caused by the vessel which has given rise to a potential liability. For incidents covered by P&I insurance, the club will need to determine the extent of its liability by investigating the circumstances. This will be conducted by the attending surveyor or another representative appointed by the club, such as a local correspondent or lawyer.

In addition to the P&I certificate of entry, the vessel should also have on board the P&I club's rule book and means of access to its current list of correspondents. In cases of urgency, Masters can save time if they contact the nearest of the club's local correspondents themselves. They should, however, be guided by company policy on this, as many owners prefer to handle notifications to their P&I club themselves.

Whenever a P&I investigation is initiated, it is vital that the Master, officers and crew members cooperate fully and openly with the investigators. Only with full knowledge of the facts of the incident can the P&I club properly protect the owner's interests. The P&I club is on your owner's side and is there to assist, not to attribute blame.

Further guidance on the actions necessary following a casualty is available in The Nautical Institute's publications *Casualty Management Guidance* and *The Mariner's Role in Collecting Evidence*.

The Master's responsibilities in law

By Captain Richard Springthorpe

The Master must make decisions on all aspects of navigation and management of the vessel every day. Each decision can have significant consequences to all parties concerned in the maritime adventure. Masters must always be in a position to justify any decision that they make, whether made with outside assistance, such as the pilot, or not.

The purpose of this article is to support your understanding of how the Master's responsibilities are defined in law and how you may practically apply these responsibilities in your daily duties.

The Master's responsibility in law is stated as being a servant in law, an agent both for the principal, the shipowner, and to some extent the owner of the goods being carried. As Master you must obey the instructions of the charterer of your ship as to the employment of the vessel. You are also the commander of your crew and occupy a position of special trust with the owners. You are absolutely responsible for the safety of the ship and remain in command regardless of whether or not the ship is in charge of a pilot at any given time.

The framework

International maritime law is a vast and complex blend of statutes or common law, customary law, case law and international conventions. A Master cannot be expected to appreciate all of the implications or ramifications of compliance with each of them. You should, however, understand the legal framework within which you operate, to understand where your responsibilities have come from and how they are defined.

A Master has a duty to perform the role to the reasonable standards of the profession. This is the duty of good seamanship and will often require the Master, crew and owners of the vessel to observe the standards more prescriptively set out in international conventions and flag state laws. With regard to vessel operations and practices, the main bodies that make international laws and conventions, to which the Master must comply, are:

- the International Maritime Organization (IMO)
- the International Labour Organization (ILO)
- the World Health Organization (WHO)
- the International Telecommunication Union (ITU).

The most important standards are defined by the IMO in the following conventions:

- *Convention on the International Regulations for Preventing Collisions at Sea 1972*, (COLREGs), which sets the standards of care required by persons navigating vessels
- *International Convention for the Prevention of Pollution from Ships (MARPOL)*, which sets out the law relating to the use and disposal of marine pollutants onboard
- *International Convention for the Safety of Life at Sea (SOLAS)*, which establishes the requirements for safe construction of all types of vessels, including fire protection and lifesaving appliances, and safe operations
- *International Convention on Standards of Training, Certification and Watchkeeping (STCW)*, which sets the standards for the quality of crew, including licensing, training and practical guidelines such as hours of rest etc.
- Under SOLAS: *The International Safety Management Code (ISM Code)*, which establishes the framework for owners and managers to implement safe working practices and procedures, referred to as the Safety Management System (SMS).

Further, the Master must have a good understanding of the United Nations *Convention on the Law of the Sea 1982* (UNCLOS), which defines the interrelationship between international conventions and local laws, setting out countries' responsibilities for the use of the oceans and providing guidelines for business and management of natural resources.

In international waters the laws of the vessel's flag state apply, while in exclusive economic zones, coastal waters and ports, the Master should be aware that local laws may also be applicable. These local laws may be very different to international laws and can place significant legal burdens on the Master.

The ISM Code, if properly implemented in an SMS, can be of great assistance to Masters because companies are required to clearly define and document their authority and responsibility regarding:

- Implementation of the company's safety and environmental protection policy
- Motivation of the crew in observing that policy

- Issuing appropriate orders and instructions
- Verifying that specified requirements are observed
- Reviewing the SMS and reporting deficiencies to shore-based management.

Modern means of communication often allow instant advice and instructions from ashore, which can mean that decisions previously reserved for the Master are now made onshore. However, you must remember that you remain fully responsible for the operation of the vessel, and therefore any action taken must be reasonable and based on appropriate advice and relevant information. Masters facing civil or criminal liability are unlikely to be able to defend themselves on the basis of simply following orders. The Master must be autonomous in executing the role, meeting the standards of the profession, and act within the applicable law.

Day to day practicalities

The Master's daily job will be governed by the requirements of international law, local laws which are applicable, flag state laws and the company's own management procedures. There may also be non-mandatory industry guidelines to comply with, such as those drawn up by OCIMF and in ISGOTT for tankers.

The professional standards for a Master require the demonstration of good seamanship. This can be defined as the exercise of all reasonable care and skill in navigation, of at least ordinary care and ability in the transaction of business connected with the ship, and the constant use of patience and consideration in dealings with those under your command or entrusted to your care.

The Master must, of course, manage the operations and business of the ship, deliver a service and take care of the cargo being carried. In order to perform this responsibly and appropriately, you must be duly qualified and experienced to enable you to maintain a seaworthy vessel.

There are two key aspects of seaworthiness that arise in law (although this can vary in different states). Firstly, the ship, her crew and her equipment shall be in all respects sound and able to encounter and withstand the ordinary perils of the sea during the contemplated voyage. The second requires that the ship shall be suitable to carry the contract cargo.

Seaworthiness is judged by the standards and practices of the trade in question at least so long as those standards and practices are otherwise reasonable. This includes the owner's recruitment of the Master and crew. If they are found not to comply with standards for training and certification requirements, then the vessel is not seaworthy.

The Master is not to be judged against the most highly skilled and experienced Master, but against the threshold of the ordinary and reasonable Master. The decisions made and actions taken must be based on information reasonably available. You have a duty to obtain the most relevant information, on which to make those decisions in whatever operation is being carried out.

Section 1
The Nautical Institute on Command

Decisions and emergencies

A good example of the exercise of the Master's responsibility, authority and decision-making, is whether to enter a port that may be unsafe for the vessel, or whether to leave a port due to bad weather. While a vessel has an obligation to follow reasonable orders of employment from charterers, the Master is not required to take the vessel into an unsafe port.

The Master's obligation on receipt of an order from charterers is not one of instant obedience but of reasonable conduct. Only an unreasonable delay constitutes a refusal to obey an order. Before entering a port, you must check the depths, tides, navigation and the latest local warnings, including weather, keeping in mind the characteristics of the particular vessel, if you are to demonstrate that you have exercised reasonable care. Once in port, if bad weather is forecast, then the Master should consider additional moorings and have the engine standing by ready to depart, as well as seeking advice from the subject port as to whether to leave the berth.

Such actions by the Master are important because in the event of an accident, charterers may avoid liability even if they nominated an unsafe port, if the Master's actions were the proximate cause of resulting damage. Whereas, if you act reasonably, perhaps even though mistakenly in the situation confronting you, it may be that the owner can establish that the port was unsafe and that the incident could not have been avoided by the Master's good seamanship. That is to say, it is unlikely that your actions will be held to have been the effective cause of the damage.

Other obligations

A Master's obligation to render assistance at sea is a longstanding humanitarian maritime tradition. It is also codified in UNCLOS.

Where an incident occurs which involves marine pollutants, the Master has a statutory duty under MARPOL to report it without delay and to the fullest extent possible. You must also take whatever action you can to assist in cleaning up the pollutant and minimise the pollution and impact on the environment.

If your own vessel is placed in a difficult situation, then you must do everything in your power to make the vessel safe and to protect the crew and the environment. You can only protect the crew and environment by first making the vessel safe (so long as that is a reasonably viable option, or, if not, giving appropriate orders to abandon ship). Delaying a decision to seek salvage assistance for example, to connect a salvage tug to save the vessel from imminent danger (because owners are trying to secure a more cost-effective option) may not be considered reasonable. The Master is the only one who can accurately assess the risk to the vessel and crew.

The Master must strictly comply with the COLREGs, but in addition must also ensure that watch systems and procedures are properly enforced onboard. It is not reasonable for example to allow the lookout to rest, or perform duties such as cleaning at night or in busy waters, leaving only the watch officer on the bridge. The Master must decide, keeping safety as the main priority, how best to utilise the crew to ensure good performance onboard and ultimately to comply with company procedures and best practice guidelines.

Cargo operations and related issues are addressed in more detail on pages 47-49 and 61-63 of this book. However, it should be noted that the Master has an obligation to manage cargo operations (subject to alternative responsibilities set out in particular contracts). Cargo must be secured for all expected conditions on voyage, including bad weather and changing temperatures, supported by calculations wherever possible, and strictly according to the ship's cargo manual for that trade.

As Master you will be given considerable latitude by the courts when you find yourself in difficult situations. Provided you have acted reasonably in the face of that difficult position, then you can confidently take the action you consider appropriate at the time. In court proceedings, expert evidence will be presented to determine what a reasonable Master would do in any given situation. Such advice is used to gauge what is the standard of perfection compared with what can be ordinarily and reasonably expected of a Master.

In summary

It is the Master's responsibility in law to act reasonably and to exercise care and skill in any operation or management decision onboard. The SMS put in place by the owners/managers in accordance with the ISM Code should provide the primary framework for the Master's responsibilities and enable you as Master to perform your professional duties through implementing and enforcing the SMS onboard while considering each situation on its own facts and merits.

Section 2

Operational aspects

Handover

By Captain Ozan Dermen

A handover should cover all the important points for which the new Master becomes responsible as soon as the old one is relieved. Asking the right questions in the right order will ensure an efficient and smooth handover.

From the first minutes of the command the new Master is supposed to know everything about:

- Conditions affecting the seaworthiness of the ship
- The handling characteristics of the ship
- The ship's business
- Personnel matters
- Preparations for forthcoming events
- Details of recent near-misses or incidents.

Seaworthiness

Customary maritime law describes a seaworthy vessel as tight, staunch and strong, and in every way fitted for the voyage. For handover, it is far more efficient to look at what makes a vessel unseaworthy and concentrate on whether these conditions are present or not.

A ship is deemed to be unseaworthy if:

- She does not have the correct and valid statutory certificates
- Her navigation equipment is not ready in all respects for the intended voyage
- She does not have the correct number and certified complement of crew
- She cannot provide the working and living conditions required by the Maritime Labour Convention
- Her cargo has not been properly loaded, stowed and secured for the voyage
- She has not been supplied with sufficient bunkers, water and provisions for the voyage.

The first step of the handover is to review the certificates. As well as verifying the validity of the certificates, upcoming surveys should be also noted. However, the presence of valid certificates is not enough to make a vessel seaworthy. Her statutory equipment should also work properly. Details of any repair required to rectify a recent breakdown should be discussed and clarified with the Master signing off.

Even if the next destination is not known to the Master, the bridge should be ready to navigate in all respects. The charts and publications required should be ready to use and up to date. When the destination is known, a passage plan should be ready and reviewed by the new Master and all officers before departure. All navigation equipment should be in good order and new crew members who have just signed on should complete a familiarisation of their immediate responsibilities before departure.

If a ship is not cargoworthy she cannot be seaworthy. A cargoworthy vessel should provide secure stowage and also have the facilities to take care of her cargo during the sea passage. Properly working ventilation fans for bulk carriers, enough reefer plugs for container vessels or heating coils without any leaks for tankers are a few examples of the proper care sought by the shippers. Keeping in mind that commercial vessels are designed to earn freight, Masters taking over should familiarise themselves with the condition of the cargo handling systems as soon as possible.

A ship without enough bunkers, water or provisions is not considered fit for the intended voyage. Furthermore, certain charterparties may consider any supply attempt as a deviation once the vessel has sailed from the load port. For this reason, it is imperative that the Master signing on immediately checks the amount of bunkers, water and provisions on board. When calculating the required bunkers, pay specific attention to:

- Whether the ship has to pass through an emission control zone (ECA) or visit a port with strict emission control standards requiring the consumption of different kinds of bunkers, such as low-sulphur fuel oil or marine gas oil. If so, does the ship have enough supplies of each kind?
- Whether any extra fuel consumption is expected during the sea passage from operations such as purging, cargo heating or tank/hold washing
- Whether the ship has enough extra bunkers for unexpected events depending on the season and the route, such as bad weather or search and rescue operations.

Shiphandling

Every ship has her own manoeuvring characteristics. Even though the ship's wheel house poster displays her manoeuvring information, the outgoing Master's ship-specific experience is invaluable for the new Master. Taking into account the ever changing factors of hull cleanliness and the effects of loading on manoeuvrability, signing-on Masters should familiarise themselves with the specific handling characteristics of the new vessel as far as possible. This includes elements such as the swing of her bow when the engine is on astern, the critical rpm of her engine and her responsiveness to strong winds and engine commands etc.

Ship's business

The technical manager's agent is the main connection for the Master to the local network of the port of call. The agent knows the national rules and practices and carries out what is required by local immigration and sanitary authorities and customs. A slight error in the paperwork or a misunderstanding during its preparation may result in a delay or even a detention.

The technical manager's agent can speak to the local service providers at the port of call in their own language, which can be particularly useful during an emergency or commercial conflict. The new Master should note full contact details of the technical manager's agent and understand the status of the activities the agent is carrying out on behalf of the ship. The ship's agent may be arranging the pilot, the issue and transfer of port clearance to the ship and deliveries of various items such as emergency equipment or provisions, among other services.

The joining Master should discuss the charterer's instructions with the current Master, because it is likely that the ship has a second agent assigned by the charterer. It is equally important to have contact details for the charterer's agent, because often they will arrange most of the commercial activities required for the successful execution of the next voyage. Examples include bunkering, deliveries of cargo-related equipment or test kits, and transportation arrangements for cargo surveyors.

If available at the time of the handover, the joining Master should examine the voyage orders. A voyage order is the summary of the charterparty undertaken for the next voyage. As well as describing the intended voyage, the voyage order explains the technical requirements for cargo preparation and carriage and also the reporting procedures and commercial transactions to be followed by the Master.

Once the voyage order has been thoroughly read and understood, the new Master should assess the capabilities on board and decide if the intended voyage can be executed safely with the present resources. If replenishments are needed, arrangements for the required requisitions should immediately be discussed with the Master signing off. As breach of the voyage orders after their acceptance may cause serious economic losses for the technical manager, any unclear points should be raised for further discussion straight away.

Personnel matters

A well-equipped ship cannot be deemed seaworthy if not under the control of a well-trained crew. The new Master should ensure that the number and qualifications of the crew comply with the safe manning certificate and that all the crew have medical certificates that will remain valid until completion of the next intended voyage. The new Master should also confirm that the crew will have had enough rest at the planned departure time. A tired crew member, who has been deprived of the minimum rest granted by the regulations, cannot be considered fit for duties.

The Maritime Labour Convention 2006 entered into force in 2013 and sets out the minimum standards for a safe and healthy living environment. It also clarifies the labour relationships between the crew and the company. Under it, Masters have a statutory duty to confirm that the ship can provide the crew with the safe, healthy and decent living environment required. It is also their responsibility to clarify if there is any labour dispute between crew members and the employer. If you become aware of a dispute or a complaint, its status should be discussed with the outgoing Master while the ship is still in port. For more details see pages 81-84.

If you have the opportunity, discuss the crew's latest appraisals with the outgoing Master. If not, study these at an early stage as they will help you identify strengths and weaknesses.

Preparations for forthcoming events

The joining Master should find out if any events have been arranged for the next port of call. This is especially important if annual surveys need preparation, as they usually require numerous service engineer visits and equipment landings. Examples of such events include ISM documentation that was delayed as a result of a recent busy schedule, voyage summaries of previous voyages having fallen due on arrival or anti-piracy precautions to be taken for the next voyage.

Recent near-misses or accidents

A quick review of recent near-misses and the precautions adopted subsequently can help the new Master prioritise tasks to improve safety during the first days of command.

In spite of its critical role, handover usually takes place over a short period. Before both the leaving and incoming Master sign the official log book, the new Master should inspect past entries to ensure that all the official notifications have been recorded and that nothing is outstanding. Taking into account the potential fatigue of the Master signing off and the stress of Master signing on, this should be treated as a vital part of any handover. Covering all the essential points for crew safety and for a smooth commercial operation demands good planning.

ISM expectations

By Kevin Slade

Much has changed since the first edition of this book was first published in 1986. That edition made no reference to communicating with the office ashore; the only options then were letter, telegram and telex. Nor did the book mention the need to adhere to a safety management system (SMS), because the International Safety Management (ISM) Code was not yet in existence.

The 21st century has brought increased pressures and more concentrated workloads, but it has also provided us with valuable tools, including computer systems, networks and instant communication. These tools should be used to create a culture that facilitates good management principles – led by the Master.

The ISM Code requires the ship to have an approved safety management system, which in turn must be based on sound safety management principles. A good SMS defines the roles and responsibilities of all on board. In particular, it sets down what is expected from

Masters by way of governance, compliance, reporting and administration, and states their responsibilities and ultimate authority. A significant and welcome factor is that a modern SMS is also designed to allow Masters to take command of any situation with the full and immediate support of a strong shore organisation, freeing up Masters to concentrate on the issue at hand. This is teamwork backed up by single-point communication.

The chief responsibility of the Master is to ensure the safety of the crew and the vessel. You achieve this by adhering to a safety management system that is properly and safely applied by making full use of the tools supplied – people and systems. This can only be accomplished through the application of well-scoped procedures and policies, operating within a properly structured management system with defined roles for all on board.

A SMS must be practical and cannot be remote. It is a live system – constantly audited, reviewed and amended. The Master supplies feedback and advises whenever opportunities for change are identified. There will always be room for improvement, so continuous feedback is necessary and expected in order to build upon good practice. In support of the Master, the SMS undergoes continual reassessment by competent shore management, acting as the 'eyes and ears' of the Master on shore.

The vessel's owners or managers play a major role in taking the strategic view of the Master's position on board, providing the modern tools and networks to do the job but also carrying out necessary support and guidance vital to operating a well-run ship.

A key role of the Master is that of a mentor. Mentoring must be built into the SMS and actively encouraged as part of the company ethos. The SMS defines a structure or framework that allows mentoring to take place, establishes policies by which mentoring can grow and promotes cultural awareness to enable mentoring to flourish (see pages 91-93).

Shore management expects a Master:

- Not to be afraid to take advice
- To practise due diligence and honesty in reporting – 'as is' and not 'as expected'
- To have an awareness of, and a respect for, different cultures, both on the ship and ashore. Cultural awareness is so important on board the modern ship, where multicultural crews are linked by a common language
- To welcome audits and checklists and see them as a learning process, aide memoire and a challenge!
- To be a role model to the officers, encouraging and participating in training, drills, debriefs and constant improvement
- To make effective use of time zones, to be aware of shore office operating hours and to send emails for office opening time, not closing time
- To prioritise emails, remembering that some do not have to be answered immediately, particularly when the sender and ship are in different time zones; to be aware of the hours of rest requirements
- Not to allow an 'us and them' culture to develop, recognising that many shore managers have previous experience as senior officers
- To encourage and monitor onboard use of the working language of the SMS.

The Master will inevitably communicate with people who are ignorant of shipping vocabulary. Ballast and bunkers, laycan and hove to are some of the commonly used terms that are likely to bewilder those without maritime knowledge. The Master needs to understand that a seafaring background is not necessary for many functions ashore that support ship management.

Some practical tips on communicating with shore staff:

- To non-specialists, seafaring terms can be misleading. They can easily be replaced by simple language understood by all. Be patient and take time to explain the situation in layman's terms
- Never take delivery of sent emails for granted; do not to be afraid to follow up if an answer does not arrive in a reasonable time. Simply ask for an update
- Politeness costs nothing. A bad-tempered email, especially one affecting cultural sensitivities, can destroy a good relationship
- The person receiving the email may not share the writer's language, so avoid colloquialisms, which may not convey the sense the writer intended
- Sarcasm simply does not work
- Do not be ambiguous: keep content clear and understand that language can be a barrier. Often a telephone call can clear up a misunderstanding and avoid creating a long and unproductive string of emails
- Where possible, reflect on your response and make use of time zones so that your email still arrives by office opening
- Take time to look at a situation from the shore viewpoint – is there a possibility of misinterpretation?
- And finally, show respect to gain respect – respect costs nothing!

Fatigue management

By Michelle Grech

Today's maritime industry requires continuous operational capability, which has led to awareness of the dangers of fatigue. This occurs when a person's physical or mental limits are reached and they can no longer function within normal safe boundaries. Fatigue does not discriminate and can affect anyone, irrespective of training, knowledge, competency or rank. A particularly dangerous aspect of fatigue is that individuals are often unable to recognise its onset and do not realise their performance is compromised.

Fatigue can manifest itself in several ways, such as constant yawning, lack of concentration, reduced attention, reduced communication, slow responses, irritability, low motivation and difficulty keeping eyes open. On the bridge, nodding off while on watch happens more often than is reported and can be very hazardous. An individual experiencing this so-called micro-sleep is usually unaware that it has occurred and may continue to perform simple tasks, possibly in auto mode. However, when carrying out safety-critical tasks, such as navigating the ship through congested waters, the consequences of micro-sleep may be severe. Accidents like the *Shen Neng 1* grounding illustrate the dangers of fatigue at sea. The report is free and available online from the Australian Transport Safety Bureau.

You and your crew are likely to experience fatigue during the course of your shipboard work. It is important to understand the causes of fatigue and how to manage it, as it can seriously affect harmony, safety and well-being aboard ship.

The two most important processes that influence the onset of fatigue are sleep debt and time of day.

If you have less sleep than you need you build up a sleep debt. Each successive day or night that you are deprived of sufficient sleep adds to that sleep debt, accelerating the onset of fatigue. To sustain continuous operations on ships, some working and shift arrangements are employed that may be less than ideal, exposing you and the crew to some level of sleep debt. Members of the crew who have to sleep during the day will build up a sleep debt because daytime sleep quality is usually poorer than at night. Any crew members who suffer from sleep-related disorders, such as snoring and sleep apnoea (a breathing problem) or other medical conditions are more likely to accumulate sleep debt.

The time of day at which work takes place is important in determining alertness levels. An individual's circadian rhythm (internal body clock) is on a 24-hour cycle and helps regulate sleep time. The circadian rhythm dips at night, causing certain biological functions to slow down and increasing the drive for sleep. The dip is at its lowest between 03.00 and 05.00. This period is known as the window of circadian low and is the time when we are least mentally alert. Studies show that errors are more likely during this period. Anyone who has built up sleep debt and is working during the window of circadian low has to cope with a particularly hazardous combination, and the risk of errors increases substantially.

Once the window of circadian low has passed, alertness levels begin to rise steadily and continue to rise during the day, generally peaking in the early evening hours, around 16.00-18.00. A slight dip during the day – commonly referred to as the post-lunch dip – occurs around 14.00-15.00, which can also increase the risk of fatigue.

The circadian rhythm makes it easy to initiate sleep at night, while during daylight hours staying awake and being alert feels natural. Crew members who need to work at night have to stay awake when the drive for sleep is at its highest and then must try to sleep during the day when they are naturally predisposed to feeling alert (except during the post-lunch dip). Daytime sleep is shorter and intermittent, often resulting in sleep debt. Long exposure to this sleep pattern can cause more serious chronic fatigue health issues. Be aware that people who engage in night work never adjust and continue to experience a drive for sleep during the night and poor sleep quality in the daytime.

Working at sea involves other factors that increase the risk of fatigue. These include long working hours, operational demands, motion, noise and vibration. In addition, the crew not only work but also sleep at their workplace. Sixteen hours of wakefulness – including both work and non-work time – can impair performance.

In general, hours of work and rest are regulated by the IMO's Standards of Training, Certification and Watchkeeping for Seafarers (STCW) and the International Labour Organization's Maritime Labour Convention (MLC). Although STCW and MLC set a global minimum standard across the industry for work and rest hour requirements, they provide only one layer of defence and are not adequate to manage all the risks of fatigue at sea. This is because they are a one-size-fits-all approach and do not consider all of the issues discussed here.

Responsibility for managing the risks of fatigue goes beyond the ship and its command. The shipowner or technical manager also has a duty to ensure that you are supported to manage this risk effectively. However, the reality is that sometimes you will have no choice but to operate with limited crew and resources and will have to manage operations and watch schedules that will tend to increase the risk of fatigue.

The single most effective strategy for reducing the risk of fatigue is ensuring that everybody on board, including you, obtains sufficient sleep. Rest can relieve tired muscles, but the brain can only recover when provided with adequate good quality sleep. Although the amount of sleep needed to ensure optimal performance varies between individuals, it is generally considered that between seven and nine hours of continuous sleep per night is required. Recent studies suggest that split sleep can also offer similar recovery benefits so long as the total amount of sleep is maintained over a 24 hour period. The shipboard environment is conducive to split sleep, but the priority is for all crew members to have sufficient time for continuous uninterrupted sleep.

The following are some strategies that you can adopt to manage shipboard fatigue:

- Be wary of the window of circadian low (03.00-05.00) and the post-lunch dip (14.00-15.00)
- Where possible, avoid assigning critical tasks during the window of circadian low. When it is essential to navigate through a hazardous or sensitive area during the window of circadian low, be especially vigilant and self-aware. Educate the crew about the issues of working during these times and how it can adversely affect them
- Request crew members to monitor each other and be extra vigilant, especially during the window of circadian low
- Watch for signs and symptoms of fatigue and take appropriate action
- Avoid sitting down during the window of circadian low
- Try to stretch and walk around, and, if the access design of the wheelhouse allows, periodically walk outside for some fresh air
- Where possible, keep the lights bright at night and the temperature cool
- Where possible, place crew working night shifts in cabins far away from noise and disturbances. Ensure that their cabins are dark and windows blocked out during the day when they have to rest and sleep

- Make sure you are well rested, and, if required to work during the window of circadian low, take a brief nap at a convenient time
- Consume caffeine carefully and avoid it before bedtime, as it can make it more difficult to fall asleep
- When working in warm temperatures, ensure that you and your crew get plenty of rest breaks
- When investigating and/or reporting incidents, note whether fatigue was a safety issue.

State of the ship

State of the ship: cargo

By Ian MacLean

Disputes relating to cargo are among the most costly that shipowners can face. The extent to which cargo claims can be successfully defended will depend largely on the shipboard records and the actions taken by the vessel at the time of loading, stowage, carriage and discharge. Of particular importance is the Master's management of the commercial documents described in this article.

The golden rules

Know who to approach for commercial advice

You will often be faced with making decisions that may involve a commercial risk for the owner of the vessel. Where there is any doubt in your mind (as in the examples given in this article), contact the owner's or technical manager's representative for advice. If there are communication problems, take the advice of the local P&I club correspondent. These are the sources to go to whenever you need commercial advice.

Know the voyage instructions

If it is available, request a copy of the charterparty. However, charterparties are often agreed through an exchange of emails where terms are negotiated but no final charterparty is produced. You should make yourself familiar with the requirements of the voyage instructions, which should be in writing. Seek advice if (a) you are unsure about the instructions, (b) after checking with the chief engineer and chief officer, you are concerned that the vessel cannot perform as required by the voyage instructions, or (c) you are given instructions by any party (including the charterer) that contradict the charterparty or written voyage instructions.

Recognise the alarm bells

These should ring when a party pressurises you to sign a document that you do not understand, is illegible or that leads you to doubt its contents and accuracy.

Be especially wary if you are told, for example, that (a) it's the custom to sign these documents here, (b) we will give you a letter of indemnity meaning that neither you or the owners will be liable, (c) you will be in breach of contract or committing a local criminal offence if you don't sign, or (d) the ship will be delayed if you don't sign. In such cases, always seek advice. Never sign documents that refer to terms and conditions that you have not seen.

Sign for receipt
If you are unsure about the accuracy of a document presented to you for signature, insert: 'Signed without prejudice, without any admission and for receipt only.' If there are parts of the document that could be added to after you have signed it – for example, a draft survey that only covers the pre-load figures – draw a line across the uncompleted section so that it cannot be filled in after you have signed it.

Keep copies of documents
You should keep copies of all documents you are asked to sign, including the reverse side if it includes pre-printed terms.

Record instructions in writing
When you are given verbal instructions, consider sending an email to the person who gave you the instructions and state what you understood the instructions to be and what you intend to do. This serves as a written record of what was agreed and gives the recipient the opportunity to correct any misunderstandings.

Understand the value of contemporaneous documents
Documents created as soon as something happens are given great weight by the courts. A record of a hose test of a hatch cover made at the time the test was carried out is likely to persuade the court that the testing was done as recorded. Most operations will be covered by the vessel's safety management system. Make sure that the checklists that record activities are properly completed. They should not be devalued by ticking everything as OK even when an item is deficient or not present.

Preparing for cargo
The vessel has an obligation to provide spaces that are fit and safe for the reception, carriage and preservation of cargo. If there is no checklist entry for activities such as cleaning and inspecting holds, testing bilge wells, hose testing hatches, testing heating coils etc, make sure an entry is put in the log book. The best way to prove that the job was done is to have to a record created as soon as the job is completed.

Notice of readiness (NOR)
The voyage instructions will tell you when to tender the NOR. It is important to follow the instructions concerning the person it is to be addressed to, the method of delivery and the timing of delivery. An NOR may not be valid if the vessel is not in all respects ready to work cargo. This could include cleanliness of the holds, operation of hatch covers, status of generators necessary for cargo work and availability of cranes and derricks.

If you are in doubt, seek advice. If there is a question over whether the vessel is ready or not, or in the correct location, it may be possible to issue a first NOR immediately and a second NOR later on. Seek advice as to the wording for second and subsequent NORs.

Bills of lading

You should ensure that the quantity and quality of the cargo is as described on the bill. Seek advice if:

- You believe either the quantity or the description of the goods on the bill of lading to be incorrect and you are being pressurised to sign. Note that even if the charterparty or voyage instructions states that the Master must sign the bills 'as presented', under English law you cannot be required to sign bills that are factually incorrect.
- You are offered a letter of indemnity in return for signing a bill that is factually incorrect. Under English law, if a letter of indemnity is issued where the bill is clearly wrong, the letter may be unenforceable on the grounds of fraudulent misrepresentation.
- You cannot determine the number, quantity or weight of the cargo loaded and you are not permitted to clause the bill, number unknown, quantity unknown or weight unknown.
- The cargo and/or packaging is damaged, abnormal, defective or inadequate and you are not permitted to clause the bill with appropriate wording such as: ten steel coils damaged by handling gear or order and condition unknown.
- The ship cannot determine by reasonable inspection the cargo marks or serial numbers of seals and these are referred to in the bills and you are not permitted to remove the reference or to clause the bill 'cargo marks unknown'.
- The date on the bill is not the date the cargo was loaded or that loading was completed. (Insurers may exclude claims in some circumstances for incorrectly dated bills.)

Discharge port

As a general rule, cargo should only be delivered to the party that presents the bill of lading.

On occasion, you may be asked to deliver a cargo against a letter of indemnity. This letter will promise to compensate the owner should it later turn out that the cargo was delivered to the wrong party. Although this letter of indemnity would be enforceable in court, it would be of no value if the person signing it had no money to pay compensation or could not be found. Insurers will not provide cover for losses arising from accepting such an indemnity.

As Master, you should never be required to decide whether to take the risk of accepting a letter of indemnity, so you should seek advice before signing one if you do not have written instructions.

Section 2
The Nautical Institute on Command

State of the ship: certification

By Kenny Crawford

Certification requirements for ships have grown rapidly with the adoption of many new international conventions. In the past 20 years we have seen the introduction of the ISM Code, ISPS Code, Marpol Annex VI and the Maritime Labour Convention, among others. If certification is not monitored and controlled, there is a risk of the vessel being detained, and, in the event of an incident, insurance might be invalidated. The company should have a system in place to monitor certification for all its ships, but it remains the Master's responsibility to ensure that the ship is appropriately certificated.

Certification can be separated into two categories: ship certificates and seafarer certificates.

Ship certificates

Numerous certificates are issued to a ship in the course of its life. This starts with the handing over in the newbuild folder of the various test certificates for anchor chains and lifeboat launching appliances and type approval certificates for oily water separators. These certificates generally do not have an expiry date and do not need subsequent survey endorsement.

A ship must be registered with a flag state and hold an appropriate certificate of registration. This effectively gives the ship an identity and records its main details. A classification society will issue a certificate of class, setting out the ship's status with regard to design, construction and maintenance of the hull and machinery.

Various statutory certificates are issued by the state of registry or, on its behalf, by a recognised organisation – sometimes the ship's classification society – which is authorised to conduct the surveys and issue the flag state's certificates.

The IMO's comprehensive *List of certificates and documents required to be carried on board ships* is updated at intervals and you should refer to the latest version when checking your ship's certificates. A table giving details of the certificates that a vessel may be expected to possess is available at www.nautinst.org/command. It should be remembered that not every certificate will be relevant to every vessel – the passenger ship safety certificate is not needed on a cargo ship, for example.

Most certificates have an expiry date. IMO's harmonised system of survey and certification is an attempt to align the various certificates' expiry dates. Therefore, particularly when taking over command, the Master should be aware of the certification requirements for the type of vessel and ensure that all the necessary certificates are in date. The certificates may have been issued for five years, but the main class and statutory certificates are subject to annual and intermediate surveys, so the Master needs to know when the ship is in the survey window (usually plus or minus three months from the anniversary of the expiry date). See pages 39-42.

The ship may be carrying interim certification instead of full-term certificates. This may be on completion of renewal surveys or because there has been a change of flag or of management. Should this be the case, the new Master must ensure that they are not expired, or due to expire during the voyage. You must be satisfied that the company will have the full-term certificates delivered to the ship before the interim certificates expire. Some of the requirements may be subject to exemptions in certain circumstances. An exemption certificate issued for a particular item also needs to be noted and monitored.

Seafarer certification

In an ideal world there would be only one type of seafarer certificate for all ranks on all types of ship. The reality is completely different. Masters and officers must hold a valid certificate of competency issued by the flag state, appropriate for the rank served. Where the flag state of the issuing authority differs from that of the vessel, the STCW Convention permits the certificate to be endorsed, which attests the recognition of that certificate. Alternatively, the vessel's flag state can issue a certificate of equivalent competency (CEC), which meets the same need as an endorsement attesting recognition. Crew members should have the appropriate certificate of proficiency for their rank, if applicable.

The ISM Code places the responsibility on the company and the Master to ensure that each ship is manned with appropriately qualified and medically fit seafarers. Each seafarer must have a valid medical certificate that shows any limitations, such as not fit for lookout duties or a limitation on trading areas. Expiry dates should be monitored. If a medical certificate is due to expire, arrangements should be made for an approved medical practitioner in a convenient port to conduct a medical examination and issue a certificate before the expiry date.

The Master should be aware of any ship type endorsement requirements, such as tankers and passenger ships, and ensure that each seafarer has the appropriate and relevant certification. The crew list should be checked to confirm that there are sufficient seafarers to meet the requirements of the minimum safe manning document and an assessment carried out to ensure that each seafarer is appropriately certificated.

Port state control

Inspectors from port states have the right to board ships in their ports at any time. The control provisions contained in the conventions focus the inspection on the validity of the certificates. Under SOLAS Chapter I Regulation 19: *"Such certificates, if valid, shall be accepted unless there are clear grounds for believing that the condition of the ship and its equipment does not correspond substantially with the particulars of any of the certificates or that the ship and its equipment are not in compliance with the conditions of regulation 11(a) and (b)"* (Maintenance of conditions after survey).

A port state control officer (PSCO) is certain to check the certificates. The manner in which the certificates are presented can sometimes provide the PSCO with an indication of how the vessel is being run. For example, if the certificates are presented quickly and in good order, this suggests the documentation is under control, providing a favourable impression of the way the ship is being operated and maintained. If the ship has to be searched for the certificates and it takes hours to present them, the PSCO will not be impressed. See pages 28-30 for more details on port state control inspections. An example of a spreadsheet for monitoring statutory and seafarer certificates is to be found at www.nautinst.org/command

Use a program such as Excel to alert you to pending survey and expiry dates.

Conclusion

Certification of ships and seafarers is not easy to manage. The modern Master needs good administration skills to keep track of the many certificates required, those actually onboard, their expiry dates and survey due dates. The company has a duty to support the Master, but in reality the Master is responsible for the ship's operation. Time and effort should be put aside, especially during handover, to go through each certificate methodically to ensure that it is correct and valid.

State of the ship: surveys

By Walter Vervloesem

Inspections and surveys should verify that ships meet rules, regulations and industry standards. They are always a challenge for the ship's Master and the surveyor. In this article, we will focus on the soft skills which can help to make ship inspections successful.

Ship inspections: the challenge

Inspections are not always welcomed by ship or shore staff because they coincide with the vessel's hectic port schedule. Staff will be stressed wondering what will happen if the inspection is not passed, knowing consequences could include delays, arguments, off-hire, detentions or blacklisting.

Planning and organising the survey is important for Master and surveyor. The Master will have to deal with the ship's operational requirements including cargo operations, delivery of stores, bunkering and crew changes. The surveyor will need to schedule the inspection to fit in with the ship's crew work and rest hours while avoiding disruption. It's a tight time frame.

Ship inspections: the scope

Ship inspections are carried out for a range of principals: class, flag state, P&I, H&M, industry majors, financial institutions and potential buyers. Sometimes inspections cover only part of a vessel, such as the ballast tanks, and sometimes a specific type of survey is requested (such as an empty tank survey). Ship inspections also complement audits including those undertaken for ISM, ISPS or MLC. They can take one or more days. Generally surveyors are required to address:

- Certificates
- Management and procedures
- Crew
- Safety and PPE
- Accommodation and catering
- FFE
- LSA
- Environmental management
- Navigation
- Communication
- Structure
- Machinery
- Cargo related and ship specific equipment/procedures.

Ship inspections: preparation

Preparing for the survey is the key to success and here are some hints to help it go smoothly. Establish lines of communication between the ship, the surveyor and the company as soon as possible and learn about the purpose and scope of the survey.

Consider making a surveyor's file to include:

- Ship's particulars
- Certificate status record
- GA plan
- Brief description of the ship type, operational and cargo related features.

Ensure it contains valid general (not confidential) information and make several copies. Ask surveyors to select the relevant information from these documents so work can start while the Master continues with other duties.

Of course the inspector will frequently need more information, and the Master should save time by being familiar with the ship's library and filing system to rapidly find any documents or manuals required. Encourage junior officers to assist in preparing these files and documents. Check that the ship's photocopier is operational and with spare paper. Have a crewmember stand by to make copies.

Expect the survey to take several hours or even days and consider accommodation and catering for the surveyor. Hospitality is important, but sometimes overlooked.

Testing

Find out in advance which tests will be carried out. Give equipment a check to ensure the relevant crew is familiar with its operation and that it is working properly. Equipment failure during survey leads to delays, repairs and retesting. It also leaves a bad impression.

Inspection

Avoid night inspections and observe work and rest hours. Ensure that you and the crew in charge are fully aware of the requirements for permits to work (PTW). Equipment to be inspected needs to be available, operational and calibrated.

Oxygen and gas analysers need to be calibrated and to be correctly used by the persons in charge. Do not forget to rig railings and ropes, or post warnings, where necessary. Safety gear such as safety belts, goggles, ear muffs, breathing apparatus and stretchers must be readily available, clean and in order. Extra flashlights are useful. There should be enough light for safety and electrical connections must be safe.

Pre-inspection meeting

If you need to delegate an officer to accompany the surveyor, make sure that they have been properly briefed on the scope and purpose of the survey, what to do or say, and the survey schedule. Crew must inform you if the surveyor requests something that is dangerous or beyond the scope of the inspection.

Your company should notify you in advance of an inspection and tell you who is expected on board, in which port, what type of survey will be carried out and when. Preferably, a superintendent should board the vessel to liaise with the surveyor, but as Master you are responsible for its outcome.

Ship inspections: the opening meeting

All this preparation will make it clear to the surveyor that everything is under control. The opening meeting is important as now you will learn about the detailed scope of the survey and the standards against which items will be checked and tested.

Do:

- Look after surveyors
- Liaise with your company if a surveyor presents a waiver letter
- Seek advice if in doubt about the inspection.
- Ask for clarification when necessary
- Appoint the most suitable crewmember to accompany the surveyor
- Allow extra time for the unexpected
- Always act professionally
- Invite surveyors for meals – check dietary requirements
- Ask the surveyor to point out defects to the accompanying crew member, so that notes can be taken.

Don't:

- Arrange for different surveyors to witness the same test as they will have different priorities and requirements
- Try to impress surveyors by talking about your experiences; it can have an adverse effect
- Tell surveyors they will not find defects on board your ship.

Ship inspections: the survey

During the survey, expect the unexpected and try to reorganise it if circumstances demand.

Do:

- Conduct tests and inspections only when it is safe
- Ensure that tests are carried out by qualified crewmembers
- Provide master keys for areas to be inspected
- Ensure effective communication between the accompanying crewmember and surveyor
- Insist that PPE is used and safety precautions observed
- Warn the surveyor about dangers or obstacles
- Keep in regular contact with the accompanying crewmember using portable VHF.

Don't:

- Allow surveyors to handle shipboard equipment themselves
- Argue over defects (but make a note and discuss matters during the closing meeting)
- Try to quickly fix defects during the inspection (good repairs cannot be done quickly and quick repairs are not good)
- Continuously ask about the inspection's progress.

Ship inspections: the closing meeting

At the end of the survey, it is good practice to hold a closing meeting. In many cases this will be held when the ship is due to depart and at the end of a long and hectic day.

Do:

- Allow time for a closing meeting with the necessary heads of department
- Allow the surveyor time to prepare a preliminary report
- Ask for a numbered list of findings
- Give the surveyor the time to explain remarks or deficiencies
- Accept defects, but be sure you understand why they are considered so
- Accept that anomalies, discrepancies and defects have to be reported
- Refrain from voicing opinions
- Keep discussions about remarks short and to the point
- React in a clear, decisive but non-aggressive manner
- Recognise that remarks might indicate flaws in the company or onboard system that need addressing
- Take immediate action if a surveyor's behaviour is unacceptable or dangerous.

You will have to accept that defects do not necessarily follow strict rules. In many cases, a 'no' answer on the surveyor's inspection form will trigger a defect. A lack of good seamanship and poor due diligence principles may in some cases also be considered deficiencies.

Don't:
- Threaten the surveyor – it still happens!
- Try to bribe the surveyor – it still happens too
- Put an unknown USB stick from the surveyor in the ship's computer without confirming it is virus free
- Ask surveyors (except class or manufacturer's surveyors) for repair specifications
- Give excuses about the ship's age or wear and tear. A ship should always be sea and cargo worthy
- Say that a defect is unimportant
- React immediately or violently to what a surveyor says.

Signing of survey reports/defect lists

After the closing meeting, a deficiency list will normally be issued and presented to the Master: it is best to sign this 'for receipt only'. If you decide to add a comment ensure it is correct, justified and to the point and does not incriminate you or the company. It is best to delay making your own comments until you have taken time to consider the report. Prepare necessary evidence including photographs, short videos, repair reports, spare part order lists, test reports and extracts from the ship's manuals/SMS to prove that corrective action has been taken.

Lessons learned

After the inspection, and perhaps when back at sea, sit back and evaluate the results. Use information on deficiencies as a basis to revise the SMS and improve the maintenance plan. Remember that correcting defects will improve the ship's condition and/or operation.

Do not limit corrective action to rectifying the problem alone, but try to find the root cause – which may be a lack of familiarisation, knowledge or training – so any recurrence can be avoided.

Section 3

Day to day management – operations

Shiphandling

By Captain Trevor Bailey, Technical Editor

When you have to put the ship alongside a berth, or even another vessel, there is bound to be something inside you that says: 'Wait a minute, this is a planned collision! This goes against all my training to avoid collisions at all costs.' Take your time, exercise patience and care and you will be able to do it. In all shiphandling manoeuvres, do not be rushed – you are more likely to make mistakes that way. And I do know about that!

It is a recommendation in STCW 95 Section B/V (a) that:

Before initially assuming command on large ships and ships with unusual manoeuvring characteristics, the prospective Master should have sufficient and appropriate general experience as Master or chief officer, and either

a. Have sufficient and appropriate experience manoeuvring the same ship under supervision or in manoeuvring a ship having similar manoeuvring characteristics; or

b. Have attended an approved shiphandling simulator course on an installation capable of simulating the manoeuvring characteristics of such a ship.

It is normal practice for the Master to personally carry out the shiphandling of the vessel, perhaps with the advice of the pilot, but remember the responsibility for whatever happens during a manoeuvre remains with you, the Master.

If you have spent the majority of your time as chief officer on the fo'c'sle head during port approaches and departures, now that you are the Master, shiphandling may come as a complete surprise and a total nightmare! Masters who do not have the necessary skills, knowledge or experience to challenge or counter pilots' actions may find themselves entirely reliant on their advice and instructions.

Hopefully you will have spent time on the bridge with the Master for all arrivals and departures and taking the con with the pilot. If so, you should have gained the confidence and competence necessary to command the vessel during the most critical and highest risk periods of the entire voyage.

Shiphandling requires the exercise and application of considerable degrees of skill in recognising the balance required between the levels of power demanded of the main engines and controllable pitch propellers; the amount of rudder angle to apply and the judicious use of the bow thrusters, as appropriate. A very good sense of situational awareness is essential in order to achieve the satisfactory outcome of the manoeuvre without danger or damage to the vessel, those onboard and the shore infrastructure.

Shiphandling needs practice and the best way to achieve that is through a structured training programme. It is normal to start with departures in good weather conditions, followed by arrivals in similar conditions and to then progress to handling the vessel in more challenging weather conditions – but always under supervision. The decision to allow the chief officer to undertake any manoeuvre remains with the Master at all times.

You may be reading this as an aspiring or new Master, who may or may not have shiphandling experience, or as an experienced Master wanting guidance on implementing a structured training programme in accordance with your company's guidelines. Whatever stage you are at in your career, a structured shiphandling training programme will ensure that you, and those you are mentoring under your command, will have the necessary knowledge and practical experience of handling the vessel in all circumstances.

The training programme should consist of a combination of both theoretical and practical tasks, carried out in a variety of circumstances and conditions, by day and by night. The aim is to develop a portfolio of experience that may be taken into account during appointment and promotion interviews. The practical shiphandling programme should take account of the different handling characteristics of the specific ship and progress from basic to more complex operations.

The following suggestions are intended as a guideline and Masters should bear these in mind when developing a training schedule for the chief officer:

- Open water operations
- Picking up and dropping off pilots
- Anchoring (open and congested waters)
- Unberthing (without and with tugs)
- Berthing (without and with tugs).

When deciding to allow the chief officer to undertake the shiphandling of the vessel the Master should consider, for instance:

- The difficulty of the operation, arrangement of berth and UKC
- The condition of the vessel, her engines, steering and equipment
- External factors such as weather (wind, current, tidal) and other shipping in the area
- Shiphandling experience of the chief officer on the current and other ships
- Pilot's opinion or objections
- Whether tugs are being used or not
- Hours of rest of the chief officer or recent work load. Training may not be ideal if scheduled at the end of a problematic and complex discharge or just before tank cleaning operations.

Some adjustment of watchkeeping arrangements and working hours and mooring stations may be needed to allow the chief officer the opportunity to carry out practical training.

Whenever the chief officer is carrying out the shiphandling of the vessel, all members of the bridge team must be aware of this. Where this takes place under pilotage, the Master and the chief officer must obtain the pilot's agreement before any manoeuvres. The chief officer should be expected to explain his intended procedure before the operation to the pilot and the Master and will not be considered to have been responsible for berthing or unberthing by simply following a pilot's orders.

Assessment

The Master should supervise and monitor the chief officer and assess that the chief officer can clearly demonstrate an appreciation of the requirements of navigating and handling the vessel appropriate to the manoeuvre being undertaken.

There is no set number of manoeuvres that a chief officer is mandated to achieve. However, in accordance with OCIMF requirements, a record of not less than 30 operations is recommended before being considered for command.

Advise the chief officer that no matter how many manoeuvres they have achieved, they should take time to consider the task in advance; plan it well; discuss it with you and, above all, enjoy the manoeuvre. There is great satisfaction in handling a ship well.

Safe working practices

By Captain Nicholas Cooper

I will start this article with a quote from François Laffoucrière: *It is a temptation for young Masters to continue to do their former job instead of the new one. You need to put some distance between the two roles.*

As chief officer you can be friends with everybody on board, but as Master you must be friendly with everybody, but nobody's friend. This is a very fine but critical distinction. You have to keep that reserve and distance between you and the others, without becoming aloof. If you are too distant, officers and crew members will find it difficult to approach you and discuss potentially serious safety concerns.

You now have responsibility for everything that happens onboard your ship, and not just your particular department. You are responsible for compliance with every aspect of the safety management system, although the shipboard management team will manage its day to day operation.

You should not try to micro-manage every aspect of safety onboard, but you still have to know what is happening and what is planned in the way of maintenance and any jobs requiring a permit to work. In addition to the formal daily or weekly shipboard management meeting, you can make it a regular habit to join the chief officer on the bridge during the day, and take the chief engineer with you too. This provides a good opportunity to get to know your senior officers and allows the airing of thoughts and opinions that would be out of place in a formal meeting. You should also ensure that the ship's safety committee meets regularly, with actions and recommendations minuted and followed up promptly. See pages 76-77.

As chief officer you were running the deck department as you saw fit, but now as Master you may see things in a different light. Some practices that you formerly accepted through custom and practice may now appear sloppy or even unsafe. Examples may include inadequate use of PPE (a common cause of accidents and injuries on board); climbing a ladder to change a light bulb without a full permit to work or using a safety harness or fall arrester gear; crew entering machinery spaces without ear defenders; using power tools, especially with grinding and cutting discs, without heavy gloves and face protection and detailed tool box talks not held before enclosed space entries and freefall lifeboat launches.

Most of these safety issues can be ironed out over a mug of tea on the bridge. The trick is to make them sound as if they were the chief officer's ideas in the first place! Subtle differences in tone and dialogue can make the difference between alienating your chief officer and enrolling him as a willing participant in your schemes to improve safety on board. Banging the table and saying "I want!" will prove less effective than "We need to take a look at our PPE usage on board." You have shared the blame to a certain extent, and the chief officer is not made to look like the culprit.

Don't be afraid to seek advice from your senior officers, who in some cases may have far more experience than you do. But ask them what they think, not what they would do, and get their views and opinions first before you announce your decision.

An awkward situation that may arise is if your chief officer is much older than you and may bear a grudge. Sometimes it will seem as if everything you say or do is being sneered at by someone who thinks they know better. Instead of offering you the benefit of their knowledge and experience, they are uncommunicative and sulking at the back of the bridge. Be patient; one of you is going to sign off soon!

Shiphandling and management of the bridge team may also be challenging, particularly during port approaches and departures, and the more so if you have spent the past few years as chief officer on the fo'c'sle head at these times.

The more enlightened companies will have the chief officer on the bridge with the Master for all arrivals and departures, and taking the con with the pilot. This way Masters will have gained the necessary confidence and competence to command the vessel during the most critical periods of the entire voyage.

But pity chief officers who have been stuck on the fo'c'sle for their entire time in that rank, anchoring and mooring the vessel. On promotion to Master they will be thrown in at the deep end, and will be ill-prepared to command the vessel during arrivals and departures. This is bad management practice, but still widely prevalent, especially in the bulk carrier trades. The newly appointed Master will be more reliant than ever on the pilot and will lack the skills and knowledge to monitor, and correct if necessary, the pilot's actions.

Safety of the vessel and crew is paramount. You should be extremely cautious on approaches to and departures from ports, because the second most experienced officer, and the one who could be a valuable back-up to you on the bridge, is stuck up on the fo'c'sle and unavailable in tense or critical situations.

Never get angry, shout at or put down a member of the crew. You will be seen to have lost control of yourself and the crew member will suffer humiliation and loss of face. By creating hostility and resentment within the crew, discipline may be undermined and safety compromised.

Respect does not come automatically with promotion to Master; it has to be earned. Never drop the common courtesies and good manners that others expect of you. You don't have to say please every time you ask for something to be done, but never forget to say thank you.

The art of command is getting your officers and crew to do what you want, without apparently giving orders. This will take years to achieve, and you will achieve it only when you have complete confidence in yourself.

Safe management and delivery of cargo

By Captain Richard W A Brough

All cargoes require care and Masters are ultimately charged with that responsibility. This can be easier on single-purpose vessels such as container vessels, which make up much of the global fleet, but complications and irregularities may still arise.

Pre-loading considerations

Shippers expect consignees to receive cargoes in good condition. Consignments should be loaded and received onboard in good condition, stowed so that they cannot be affected by adjacent cargoes or the vessel itself and transported safely to discharge ports. There cargoes are received by stevedores and kept secure until receivers or their appointed agents collect it. Throughout this whole operation, the vessel must maintain its intact stability, which needs to be calculated carefully for each stage. For bulk cargoes, terminals and Masters need to exchange information and agree the sequence of operations.

Masters, acting on behalf of the owner or charterer, need to have sufficient information to carry out that duty and pre-plan the loading operation. They are likely to be assisted by a team of professionals including owners, charterers and/or cargo agents, Port Captains, cargo superintendents, cargo surveyors and shippers' representatives.

When planning the loading and voyage, Masters should make use of all available information about the cargoes to be carried. For example, cargoes that have a strong odour can taint those around them. Very dense or heavy cargoes require bottom-stowing, whereas light and fragile items require special protection. Perishable goods, metals and other cargoes can be damaged by condensation from other cargoes or the ship itself; this may be exacerbated if the voyage takes the vessel from a warm climate to a cold one, or vice versa.

Cargoes must be lashed and secured to prevent damage from cargo or vessel movement in heavy weather. To avoid weather damage to cargo, and also to ensure the safety of vessel and crew, the voyage plan may have to be adjusted from time to time.

Masters should be aware of the particular issues presented by containerised cargoes, including differences between actual and declared container weights, eccentric loading of containers, deck loading restrictions, stack heights, lashing requirements for the intended voyage conditions, plug-ins for reefer and temperature-controlled containers, out-of-gauge containers, exceptional dimensions, complexity of port rotations (especially for feeder vessels) and segregation requirements for dangerous goods.

Containers can be carried on many types of vessel along with general non-containerised or break-bulk cargoes. Break-bulk is defined as any cargo that cannot be containerised or unitised and covers such commodities as steel and iron products, machinery, casework, project and heavy cargoes and perishable goods.

Cargoes are shipped under commercial contracts and governed by numerous rules and conventions. Each item of cargo shipped is accompanied by a corresponding bill of lading and all of these bills are grouped together to form the cargo manifest for that voyage. The consignee can sell the cargo to a third party while the ship is on the high seas, in which case the bill of lading acts as proof that the purchaser owns the cargo.

A bank may provide finance for the cargo through a letter of credit. It is generally accepted that such documents are 'clean', ie there are no clauses attached to the bills relating to cargo damage. This can present Masters with difficulty if there is evidence of damage to the cargo before loading, which may be the case for steel and other metal cargoes, for example. The ship's duty officers are generally charged with checking for cargo damage incurred during loading.

All such damage should be noted, sometimes by the stevedores, who will then present an exception report to the owner's or charterer's agent. Prudent Masters will advise the principals in good time in case there is a need to appoint a cargo surveyor to protect the owner's or charterer's interest. If cargo arrives in the discharge port showing damage that was not notified before or during loading operations then a costly claim against the carrier may result.

Sometimes it is necessary to prepare the holds to receive cargo that might otherwise cause damage to the vessel or adjacent cargoes. Certain bulk cargoes are classed as dangerous, because they are liable to spontaneous combustion or liquefaction. For information on all such cargoes, refer to the latest versions of IMO's International Maritime Dangerous Goods (IMDG) Code and International Maritime Solid Bulk Cargoes (IMSBC) Code. The IMDG Code provides guidance on segregation and stowage requirements and the IMSBC Code on hold preparation, condition monitoring and, for cargoes prone to liquefaction and shifting, transportable moisture limits. Special rules exist for grain and timber deck cargoes to protect vessel safety and stability. It is essential to adhere to all these rules.

Loading and discharging

The duty officer is responsible for ensuring that cargo is loaded and stowed according to the pre-stowage plan and that the cargo is not damaged during the process. Bespoke lifting frames may be needed for special and project cargoes requiring heavy lifts and these will need to be retained on board for discharge. Suitable dunnage may need to be placed. For voyages involving multiple port calls, each port's cargo needs to be distinguishable and stevedore and crew access points should be kept free.

Much cargo is weather-sensitive so when there is a risk of rain, hatch covers should always be kept ready to close quickly. On occasions stevedores may want to continue working and offer a letter of indemnity; however, such instruments are not universally accepted in case of a damage claim and Masters should seek advice if unsure. Such letters are sometimes offered where cargo has been damaged, and again Masters should be cautious.

Protecting cargo and the vessel

Movement of cargo can cause damage both to the cargo and to the vessel's structure. Ultimately, it can lead to loss of stability and capsize, so it is important always to refer to the ship's cargo securing manual. For special cargoes, additional measures may be required such as extra pad-eyes for lashing anchors. When undertaking hot work, such as welding pad-eyes, it is essential to protect against fire in adjoining spaces.

Sometimes specialist rigging gangs are employed to lash cargo, but it remains the Master's responsibility to ensure the lashings are adequate for the intended voyage. Some cargoes are pre-slung and the test certificates for the slings must be retained on board for examination in the discharge port. Lashings should be checked frequently during the voyage, especially if bad weather is expected or experienced. Some types of cargo need to be monitored: the temperature of self-heating cargoes should be checked, containers examined for leaks, that temperature-control equipment is working correctly etc. Cargo holds may need ventilating during the voyage to help avoid ship or cargo sweat.

In dealing with complex individual cargoes or mixes of cargoes Masters have many good sources of information and advice including IMO publications, books published by The Nautical Institute and other bodies and case studies issued by P&I clubs. Standard texts are continuously updated and many websites offer information on cargo carriage and care.

Effective drills

By Captain Sarabjit Butalia

The Master's role in preparing the ship's crew for potential emergencies has never been more challenging. Despite operational constraints, you have to take a proactive role to ensure that the crew is ready to deal with any kind of emergency. In doing so, you need to manage time effectively, taking into consideration work and rest hours, cultural diversity, social media, restrictions imposed by ports (for example not being able to lower lifeboats) and commercial pressures.

The ISM Code provides the Master with the overriding authority and responsibility to make decisions about safety and pollution prevention and to request the company's assistance when needed, including for training and drills.

Drills and training for emergency preparedness should be as realistic as possible and be planned and conducted effectively. The Master should be clear about the aims of each drill.

Scheduling of drills will be determined by:

- Statutory requirements (SOLAS, ISPS Code)
- Flag state requirements
- Coastal states' requirements
- Company procedures and contingency planning
- Others, such as improving identified weak areas.

Frequencies of emergency exercises are set down in SOLAS and flag state regulations. Drills must normally be conducted at intervals of not more than 14 days, although some flag administrations, such as Liberia and the USA, require an emergency drill at least every seven days.

Companies may require weekly or monthly drills. Planning should ensure that all expected emergencies are covered as far as possible. Different drills should be run at the same time to assist in this.

Most companies undertake ship-shore drills to evaluate the effectiveness of the ship-shore contingency plan. Drills listed in the emergency contingency plan (ECP) include grounding, hull failure, collision and loss of stability. While some of the ECP manual drills may not be required under SOLAS, they are nevertheless an essential part of emergency preparedness under the company's safety management system (SMS).

To assist familiarisation and preparedness, use can be made of the guidance and checklists provided in the vessel's ECP, the shipboard oil pollution emergency plan (SOPEP) and, for ships carrying noxious liquid substances in bulk, the shipboard marine pollution emergency plan (SMPEP).

Drills and the ISM Code

Drills provide the opportunity to test the link between ship and shore. During the drills, communication between the Master and DPA and the shore contingency team should be established and verified.

Key contacts provided to the vessel for the shore support team should be tested and verified during the drill. In some cases the DPA may not be the first point of contact, so to improve access to shore the vessel should have a 24-hour telephone number to call. The ship's contingency manual should lay down the role and responsibilities of the incident co-ordinator.

Planning, preparation and briefing

The benchmark for emergency preparedness today is more demanding than merely swinging a lifeboat or running an emergency fire pump. The Master is expected to play a dual role as mentor and facilitator during drills and should be looking at key functions such as crew performance and their familiarity with the equipment, onboard communication and record-keeping.

Exercise subjects include:

- Collision
- Containment system failure
- Excessive list
- Hull failure
- Grounding
- Pollution
- Piracy
- Tank overflow
- Explosion
- Fire (deck, machinery space, accommodation, pump room)
- Enclosed space entry and rescue
- Bomb search and threats.

For a grounding exercise the Master needs to consider:

- Assigning roles such as coastal authority, DPA etc. to various crew members
- Editing messages on Inmarsat C or email
- Selecting a suitable location – for example, the English Channel or Singapore Strait
- Using resources such as charts and publications, stability conditions and tank conditions
- Limitations governing the transfer of weights, as onboard stability computers will probably be programmed only for afloat conditions
- Using and referring to the capacity plan, shell expansion plan etc.
- Briefing the crew about the scenario and ensuring they understand it
- Ensuring that all crew members understand their roles and are fully involved. For each type of emergency the role and responsibilities of individual crew members may change.

Other aspects that need to be considered are:

- Demonstration of the use of oil pollution prevention equipment, fire hoses and breathing apparatus (BA) sets
- Testing communication between ship and shore
- Familiarisation with the use of Inmarsat C for preparing and sending messages
- Being aware that emergencies demand both management and physical skills
- Conducting briefings for drills far in advance so that crew members thoroughly understand their role and responsibilities
- Identifying learning objectives before undertaking the drill

- Establishing key performance indicators before beginning the drill, including time taken to don lifejacket and BA sets and assemble the crew
- Using open sea areas with no operational constraints
- Preparing risk assessments (RAs) for lowering lifeboats in water.

To achieve the intended learning objectives, it is essential to monitor execution of the drill. To ensure that the drill is proceeding according to the plan, the Master and team should be thoroughly involved.

Debriefing should be carried out impartially, within a no blame culture, as soon as possible after the drill. All crew members should be encouraged to share their experiences. Accurate minutes or notes should be taken.

Communication means listening, providing advice and taking charge of any problem. It is the main tool for implementing company policy effectively and ensuring a no blame culture and conformity with the SMS, including during drills. The Master plays a key role in ensuring effective communication during a drill. Communication can be responsible for the success or failure of a drill.

Make allowance for the diverse nationalities that work on the ship and ashore, for whom English may not be the primary language.

Effective communication includes:

- Listening and not talking
- Showing you are listening – looking and acting interested
- Being patient, allowing time and not interrupting.

Records, in the form of statements of fact, should be maintained. This task may be delegated to one of the officers on the bridge.

The ship's bridge remains the centre of all activities, because it contains GMDSS and ship-to-shore communication equipment. It is essential that the bridge team and other parties involved in a drill (or an actual emergency) interact promptly and precisely.

Shipboard drills are critical to maritime safety and compliance, especially in the context of current concerns about environmental protection and shipboard security. A well-trained crew, carrying out well-rehearsed emergency procedures and communicating clearly, can save lives. The Master plays a very important role in this, from ensuring preparedness for drills to successful management of emergencies.

Inspection and maintenance

By Captain Sanjay Bhasin

As Master you are responsible for all shipboard inspections and maintenance, although the duties will be delegated depending on the type of vessel, the number of crew and the SMS philosophy of the owners or managers.

Inspections and maintenance are continuous processes whether a vessel is afloat, in drydock or during refitting in a yard. The type and level of inspections and maintenance will vary depending on the prevailing operational and environmental conditions and on your budget allocation.

Maintenance on cargo vessels falls broadly under the following categories:

Crew safety

- Hull integrity
- Lifesaving and fire-fighting equipment
- Health and safety.

Operational

- Engine and machinery
- Navigational equipment
- Cargo compartments
- Equipment and machinery for loading, carriage and discharge of cargo.

Living standards

- Hygiene
- Crew comfort
- Compliance with MLC.

Pollution prevention

- Prevention equipment
- Containment equipment.

Maintenance is governed by a vessel's planned maintenance system (PMS), which enables scheduling of maintenance of all items on board at intervals recommended by the classification society, equipment manufacturers and best industry practices.

When you assume command you should familiarise yourself with the inspection and maintenance systems on board and identify:

- Forthcoming priority inspection and maintenance items (especially those relating to the safety of the crew and vessel) and any engine maintenance due that would affect the seaworthiness and performance of the vessel
- Any maintenance that would affect the cargo to be loaded or safe carriage of the cargo already on board (if the vessel is laden): check the charterparty to ensure that this will be allowed
- The availability of spares and equipment required to carry out the maintenance identified and, if necessary, the facilities in the ports that the vessel will be calling at in the next voyage(s) to obtain spares, equipment and expertise for priority maintenance.

Then verify:

- Maintenance planned for the immediate voyage
- Long-term maintenance you will have to implement and supervise during your command.

Drydocking

A vessel's maintenance schedule and pattern is altered during the drydocking period. Regular maintenance ceases and most of the maintenance is directed towards drydock related jobs. Scheduled drydockings are well planned events and so there is ample time to review what is required to be done.

Inspections in drydock are generally assisted by classification society surveyors and owners' superintendents but the Master remains responsible for all safety on board during drydock. Before drydocking the Master should ensure that the following are completed:

- The pre-drydock checklist including precise soundings of all ballast and fuel tanks, readiness of fire-fighting equipment, vessel's stability on entering and leaving the dock
- Safety committee and management meetings before and during the drydock
- The yard's permits to work (PTW). The Master and senior officers must keep track of all PTWs and make sure that they are being properly followed, especially for critical jobs like enclosed space entry and hot work. Monitor all contractors and yard staff to ensure compliance with PPE and other safety related conditions attached to PTW. All tank lids and bottom plugs removed during drydocking are correctly replaced before flooding the dock
- You should liaise closely with the chief engineer; you should both be aware of all work being done in your respective departments
- Check of vessel trim/stability, bottom plugs and watertight integrity before re-floating
- Ensure sufficient time after flooding and before leaving the dock to check vessel systems and safety.

After the docking, it may take some time to update the vessel's database with all the repairs, class certificates, renewals, upgrades to equipment, service records, condition of ballast tanks and coatings, planned maintenance, and a host of engine-related work like deflections and clearances on almost every piece of machinery. Owners should have a system for this. You may choose to delegate this task but you must follow up on this to ensure that it is completed as soon as possible.

Unexpected damage

The detection and handling of unexpected damage can be challenging and could affect the seaworthiness and safety of a vessel.

With any serious situation you must follow any guidelines given in the Safety Management Manual. Your first reaction should be to stabilise the condition of the vessel if possible, to ensure the safety of the crew and provide accurate information to the owners and shore assistance providers. External assistance will normally be needed in many situations.

Resist the temptation to engage in incident investigation at this stage, even if pressed, until you have stabilised the condition of the vessel.

Port damage occurs in a more controlled environment and external assistance is generally available. Again, attending to the crew and vessel safety is the priority as is stabilisation of the situation. As Master, you should seek the assistance of local P & I correspondents and surveyors, preferably before any intrusion by port state investigation surveys, until you are satisfied that the peril has been minimised. Their assistance and local knowledge will be invaluable to you.

As a new Master, you are responsible for the timely completion of all inspections and maintenance. You may be overawed by all your other activities and responsibilities, especially if your vessel has a frequent and rapid port rotation, but you cannot allow inspections and maintenance to slip down the list of your priorities. If you need help, do not hesitate to contact your technical managers to ensure that there is sufficient support to enable you to devote sufficient attention and time to inspections and maintenance.

Ice

By Captain Duke Snider

Interest in the polar regions is growing as the season for accessing these remote and hostile areas is lengthening. Operating ships within any ice regime requires knowledge, skills and awareness beyond those of many mariners. Multi-year ice and glacial ice are much harder than first-year ice and there is little assistance available, so mariners must be self-sufficient.

The Nautical Institute publication *Polar Ship Operations – a Practical Guide* gives practical advice to those operating these ships. This article is a brief overview and Masters are advised to refer to that book.

Operating in ice without experience or on vessels not designed to operate regularly in ice or extreme cold poses huge challenges. Experience and the design of the ship will reduce the challenges to be faced, but additional operational aspects must be considered.

Ice may be encountered throughout the year in the polar regions. Other regions are subject to winter growth of sea ice that melts away each spring, including:

- Great Lakes of North America
- Northeast coast of North America
- Baltic Sea
- Caspian Sea
- Sakhalin.

Ice conditions vary considerably between these areas, so the Master must be aware of the specific ice conditions that might be expected along the intended route.

A number of vessel- and crew-related aspects need to be considered when making preparations. The company quality or safety management system should contain policies and/or checklists for preparation for and operating in ice.

Initially, the Master must consider whether the construction and power capability of the vessel is suitable. In many areas regulations require that only ice class vessels of specific notation are permitted entry. The vessel's ice class notation should be verified to ensure it is permissible and under what conditions or limitations. Some regions may require a specific ice certificate or equivalent that indicates the vessel has been vetted and meets the local regulations for ice strengthening and capability. All jurisdictions along the route where ice may be encountered should be contacted through agents to ensure compliance with the appropriate regulatory requirements.

Once it is confirmed that the vessel is suitably constructed, classified and certified to operate in the conditions expected, the Master must ensure that insurance coverage permits the voyage or, if not, that additional coverage is obtained. Both hull and club insurers must be notified of the intended voyage and need to verify that appropriate cover is in place. The charterparty agreement for the voyage must also be reviewed to ascertain permission and to check for specific clauses related to operation in or near areas of ice.

If the vessel and insurance coverage are suitable, the Master should examine whether there is sufficient ice experience among the bridge watchkeeping officers or if it is necessary to employ a trained and experienced ice navigator. Several operating areas require that the Master, a senior bridge officer or an experienced ice navigator be part of the bridge team. Even if not required by regulation, the engagement of an experienced ice navigator may be demanded by insurance, company policy or charterparty, or may simply be a prudent decision.

While in or near ice, full-time standalone lookouts and manual helm control should be considered. Allowance should be made for reasonably frequent relief of both positions. When operating in ice the Master should also consider doubling of the OOW. In this case, the junior officer ensures the vessel remains in safe water while the senior officer maintains conduct of the vessel and manoeuvres through ice as necessary. Engine room watches may also require augmentation to ensure prompt response capability during the passage.

Overall, the Master should assess the need for more crew to meet the increased bridge watch requirements. In addition to specific ice experience, crew familiarity with cold climate work should be considered.

With the above information at hand, an initial risk assessment should be completed (see pp 77-81 in *Polar Ship Operations – a Practical Guide*).

Managing risk

Operating in extreme cold or ice conditions brings with it additional risks, which should be factored into the risk assessment. Extra mitigation measures may be required to obtain acceptable risk outcomes.

The effect of cold on personnel and equipment must be added to the risk register. Personnel may be unfamiliar with cold weather effects and personal protective equipment (PPE), and this lack of understanding may hinder their response to normal workday requirements and emergency response. Cold weather physiological effects and cumbersome PPE will slow crew members' movement and reaction time, and extreme cold also reduces cognitive ability. Equipment may become brittle and thus prone to failure, reducing performance significantly.

It is important to remember that the operating area may be remote from medical or repair facilities, and SAR and pollution response may be days away.

Passage planning

In addition to normal berth-to-berth planning, other factors must be considered when developing a passage plan for potential ice transit.

Ice passage planning should be completed:

- Strategically for longer-range planning
- Tactically to deal with the daily and hourly changing conditions. Comprehensive explanation of ice-related passage planning can be found in Chapter 6 of *Polar Ship Operations – a Practical Guide*.

In the strategic phase, additional information should be gathered on present and expected ice conditions along the intended route. Ice should be avoided wherever possible and routeing selected accordingly. Ice routeing assistance may be advisable at this stage.

In most areas where ice is routinely encountered regulations require vessels to check in with vessel reporting services and confirm that the vessel and its crew meet local requirements. These centres can offer routeing assistance and may direct vessels to await icebreaker support and transit in convoy.

ETAs for the expected ice edge should be determined to ensure implementation of cold climate and ice transit checklists at the appropriate time. General tactics for plan execution beyond the ice edge should be laid out.

Tactical passage planning continues once approach and entry to ice has occurred. Ice and meteorological information need to be continually updated and the strategic plan adjusted where necessary. Information accessed should now rely on daily ice charts, images and reports in conjunction with onboard visual and radar observation and input from ice service and icebreakers.

In refining the route the Master should take advantage of optimum ice conditions, looking for open water, leads and areas of reduced pressure. ETAs should be adjusted based on ice conditions, visibility and additional manoeuvring that may be required for ice.

Operational checklists are provided in Appendices 1-4 of *Polar Ship Operations – a Practical Guide*.

Operating in ice

Lookouts, doubled watches and additional personnel should be posted in advance of the expected first encounter with ice. The ice edge should be approached at a perpendicular angle of attack, initially at low speed then gradually increasing to the speed appropriate for the vessel construction and ice conditions.

The Master must consider the four basic rules of operating in ice:

- Even if very slowly, keep moving, unless directed by the icebreaker or where stopping to await daylight or icebreaker assistance
- Try to work with ice movement not against it
- Excessive speed leads to ice damage
- Know your ship's manoeuvring characteristics.

Even in relatively open water, ship damage can occur through ice being thrown against the ship by wind and seas, particularly when the vessel is running at speed.

It is always best to avoid ice if possible. A longer route around ice may be less costly in fuel and less risky to the ship than attempting to fight one's way through it. Use pack ice and its movement to advantage rather than fight against it.

When manoeuvring through ice always take into consideration the vessel's turning characteristics to avoid lateral or stern impacts when turning.

The most experienced helmsman should be at the wheel and given clear instruction on their latitude to act as necessary to maintain base course while still manoeuvring to avoid ice. Clear instructions should be provided to:

- Put rudder amidships immediately when astern movements are ordered until the vessel begins to move forward again or the OOW orders otherwise
- Ensure that any impact with an ice floe occurs on the bow with the stem post
- Avoid passing close by heavy floes to decrease possibility of impacts against ice on the side shell plate
- Avoid sharp turns in heavy ice
- Turn the rudder towards heavy ice to prevent the bow swinging inadvertently towards weaker ice and exposing side shell to the heavier ice.

Follow open water leads and areas of lower pressure where possible. Do not navigate close on to icebergs, as massive underwater projections probably extend out from the visible above-water mass of the iceberg. Running close to an iceberg may cause the iceberg to topple over and strike the vessel because of its marginal stability.

A vessel stopped in ice runs the risk of being frozen in and beset. The best strategy is to keep way on the vessel if possible. Once stuck in ice the vessel is then at the whim of the movement of the ice. Methods to free a beset vessel include rapid heeling by transferring ballast or fuel back and forth from port to starboard, or even forward to aft. Cycling ahead and astern movements can be useful. When ahead, cycle the rudder full port and starboard, but ensure the rudder is back amidships when astern movement is ordered.

Working with icebreakers

Icebreaking commanding officers have many years of experience operating in ice and in shipping support operations. Use them as valuable sources of knowledge. If either directed to icebreaker assistance or guidance or requesting it, the Master should ensure that all the information the icebreaker may request is at hand, such as vessel length, breadth, draught, ice class, power, speed, stopping distance, destination and routeing preference and the ice experience of the Master and bridge officers.

Direct VHF communication with the icebreaker must be ensured. All bridge officers need to be aware of the visual and audible signals that will be used when in convoy.

It is essential that the Master follow the direction of the icebreaker commanding officer, particularly with respect to routeing, speed and distance between vessels in the convoy.

Always keep in mind that the icebreaker or a vessel ahead may unexpectedly become stopped by ice conditions. While in convoy, full bridge control should be available and the vessel on manoeuvring fuel. Helm control must be set to manual and lookouts should be detailed to watch carefully the distance between your own ship and the ship ahead or astern to enable appropriate collision avoidance if required when vessels become stopped by ice unexpectedly.

Writing reports

By Lucy Budd

Start by asking what is the purpose of the report? For example, incident reports record what happened, in the order that it happened. They do not draw conclusions. Some reports suggest a change or request action. The minutes of a meeting do both – they report what happened and list the actions that should be taken and who should take them. See pages 76-77 for advice on minute taking.

Ask yourself:

- What is the purpose of the report?
- Who will read it?
- What do they need to know?
- Why do they need to know it?

Decide on the aim of the report and write it down in one or two sentences before you start. This ensures that the reader doesn't miss the point and helps you focus on the key points. Then, decide what information you will need to collect and organise before you begin writing.

Consider who will be reading this report. How much technical knowledge will they have? Will you need to explain the background in detail or can you give an outline? The document may be seen by people other than those it is primarily intended for, and this additional readership may not have the same level of technical knowledge. Make sure that the terms you use are consistent throughout.

Navigation and structure

The document may well be sent on beyond the original recipient. Make sure it is easy to identify. It should include:

- Your name and position
- Name of the vessel
- The date of the report
- Date of the incident, if relevant
- Title and/or brief description of the contents
- Page numbers.

If there is no standard company template for this type of report, consider creating your own, to save time and make sure nothing is missed out. Page numbers and a footer with the document name on each page help with identification if the document is printed out. Be consistent in the way you name and save your reports, as this will help if the report has to be revised from time to time or if someone will be commenting on the document. The date or version number could be included in the title, for example in the form 20150707cargoreport.doc or cargo report v1.doc.

The report is no use if the people reading it cannot find the information they are looking for. The main issues must stand out. In a long and complex report, it might be helpful to summarise conclusions or action points at the beginning of the document, in addition to listing them at the appropriate point. Use section headings as signposts to make the report easy to follow and to identify key points. Bullet points can be useful for summaries or to highlight items on a list.

The basic structure for any report should list:

- What happened
- How it happened.

The information may be needed to enable investigators to perform further analysis. However, you may need to provide that analysis yourself, answering the questions:

- Why did it happen?
- What needs to change as a result?
- Who needs to make these changes?

The report should record what happened, but although you may draw conclusions from this, be careful not to blame people. A report might say: "The rating did not carry out the check as requested." It should not say: "The incident was the rating's fault because he did not carry out the check." All incident reports, even at the flag state level, are no blame.

Writing the report

Create a rough plan before you start writing in detail, so you know what points need to be covered, and in what order. Save the document often, but no more than every five minutes or at the end of every paragraph.

Keep the wording simple and straightforward. Do not use complex language just to make it appear formal. Make sure the terms you use are consistent throughout the report. Ask yourself whether the reader is likely to understand what you have written.

Photographs are a clearer and more effective way of showing damage than written descriptions. Include a caption to identify the picture and explain what it shows.

Provide a clear list of any action points. It may be useful to include a separate bulleted list of action points at the end of the document.

Read through the report to check for errors and make sure it is as clear as possible before you send it. If possible, get somebody else to read through the report as well. It is very easy to see what you think you have written, rather than what is actually on the page.

An incident report template

An incident report is written as a record of an event – such as cargo damage, collision or fire – by someone involved in the incident or a witness to it. Many companies have standard forms for this. If not, you can create your own to simplify the report planning and make sure nothing is missed out. You can always delete those sections that are not relevant.

A template for a report about cargo damage might include:

- Identifying information – the subject, author, date and place, and intended readership
- Introduction – a short summary of the contents. If the report makes recommendations, you could also summarise them here
- Details of the vessel
- Voyage to load port. This is likely to be relevant only for cargo damage and should give enough information to establish if the previous cargo or anything that happened on that voyage was the cause
- Loading operation
- Stowage and securing
- Loaded voyage
- Discharge operation
- Cargo damage.

For other reports, establish a consistent style in your template along the lines of the cargo report format above.

The Master should keep all reports in one place where they can be found easily. This should ensure that the same incident is not referred to in more than one place. *The Mariner's Role in Collecting Evidence*, also published by The Nautical Institute, contains more detailed information on collecting and reporting information.

Running meetings and taking minutes

By Jillian Carson-Jackson and Christine Dickinson

Here are some suggestions for effective and efficient running of the standard meetings required as part of the ship's company Safety Management System (SMS). You can adapt these guidelines as appropriate.

We have all attended meetings that were not really worth the time. Hopefully you have also attended some that were well run and enjoyable. Think back to those that worked and you will probably agree that the Chair of the meeting was well prepared, identified the object and focus for the meeting, kept the discussion on the topic, set time limits for each agenda item and ensured the meeting ran to schedule.

Running meetings

For you as Chair of the meeting, preparation is key. A good place to start is the minutes of the last meeting. Most company SMS detail what meetings need to be held, when they are to be held, provide a standard agenda and detail who is to attend. As attendance is expected, you should ensure information is provided to confirm the meeting. Emails can be sent to crew and messages or notifications put on message boards.

Although meetings are generally small and relatively informal because everybody knows everybody, there can be problems onboard ships with mixed multi-national crews in ensuring full comprehension and interaction. Information and participation rarely flows from the bottom up, especially in a public forum. It is important that you, as Chair of the meeting, are sensitive to possible communication blockers and are able to address these. Arrange seating to encourage all to discuss and exchange views. Another essential that you need to develop is the ability to remain neutral throughout the meeting.

General pointers:

- Decide on the meeting objectives, expected outcomes and timing
- Check attendee availability and confirm meeting details
- Identify and prepare the meeting location
- Provide an agenda in advance and invite additional items
- Circulate the report of the last meeting, plus any action items
- Keep to topic and time

- Be aware of any barriers to comprehension and interaction
- Ensure minutes are taken to record decisions and follow-up actions.

The agenda will follow a set format, depending on the company SMS, but should include:

- Welcome
- Review/agree agenda
- Review report of last meeting/update on action items
- Agenda items depending on focus of the meeting
- Any other business (could be carried out as a round table)
- Review of key outcomes/actions arising from the meeting
- Meeting close and setting of date, time and location for next meeting.

Minutes and follow-up action

If you are chairing the meeting, appoint someone to take the minutes. Remember minutes are an official record of actions and decisions made at a meeting and they should be written in complete sentences to accurately record what took place at the meeting. Any reader not present at the meeting should be able to understand what was discussed. The names of the attendees and the Chair should be noted.

Explain that the minutes do not need to include everything that was said but should be clear, concise and make sense to someone who wasn't there. They should highlight clearly any decisions made or action items identified. Action items should be attributed to a specific person (or a person to take the lead if a small group is to address the action item), with a time frame for completion and reporting.

Keep all your notes and rough copies. Minutes are legal documents and you are obliged to produce these, if asked, as well as recall your recollections of the meeting. Try to type up the minutes as soon as you can after the meeting while everything is fresh in your memory.

As Chair, you should check the draft of the minutes, including appropriate documentation such as the SMS. Make sure you approve them before they are sent to the attendees or to the company.

Section 4

Day to day management – personnel

Relationships on board

By Dr Captain François Laffoucrière

The Master's relationship with the heads of departments is crucial to the smooth and effective running of the ship. While certain vessel types will have specialised heads of department, such as a hotel manager on a cruise liner, this article will look only at the usual deck and engine posts. The heads of these departments, the chief engineer and the chief officer, constitute the marine team headed by the Master. In an ideal world they should work as a single unit.

Relationships with chief engineers

The Master's relationship with the chief engineer may be considered both from the perspective of the normal everyday running of the ship and during an emergency.

Everyday running of the ship

Regular contact with the chief engineer is essential, so that the Master can pass accurate, up-to-date information to management ashore and also to plan maintenance activities on board and consider improvements.

Sometimes Masters are also qualified engineers; perhaps they have even served as chief engineers. This may seem an advantage for various reasons; these Masters may be more skilled at handling the throttles and better able to appreciate technical information. But it can be a handicap if they tread on the chief engineer's toes. Even if you have technical knowledge, you, as Master, are not necessarily going to be aware of the fine details of the engine room.

Paying a visit to the engine control room is usually appreciated, so long as these visits do not become too frequent or lead to the Master constantly questioning the way things are done. You can help create a smooth and collaborative relationship by asking the chief engineer to accompany you when you want to pay a visit below. You can build upon that relationship by showing interest in the department. It is often during these times that the chief engineer will open up a bit more and elaborate on the basic facts. The better the flow of information and the easier the communication between the two, the more efficiently the ship will run. This can make a big difference during an emergency.

In times of crisis

The crisis may be a main engine breakdown, a blackout or even accidental pollution. In this time of tension and anxiety, Master and chief engineer have to rely heavily on each other. You need to have trust in the chief engineer and refrain from asking too many questions. This will be easier if you have previously built a relationship of trust and taken interest in what goes on in that department.

With the chief officer

The relationship with the chief officer is focused on (a) cargo operations and the stability of the ship and (b) safety on board.

Cargo operations

Cargo handling and care are generally the chief officer's responsibility, involving preparing or checking the cargo plan and carrying out the ship's stability calculations. You should refrain from interfering too much and doing the job yourself, but you do need to review the calculation, to ensure there is no mistake. Where a young chief officer is faced with an unfamiliar or unusual situation, it is important that you, as Master, put your experience to good use through mentoring. Concentrate on helping the chief officer in difficult situations rather than getting involved in day to day operations. The art is in knowing when to step in and when to stand aside.

It is a temptation for young Masters to continue to do their former job instead of the new one. You need to put some distance between the two roles.

Safety issues

Here again, the Master must let the chief officer do the job unhindered, but the two of you need to work closely and in harmony. As Master you have overall responsibility for the ship and should show interest in the safety duties; you will also be the one to introduce new ideas and implement company policy.

Put aside some time to work with the chief officer to develop safety drill scenarios. There is a double advantage, as this will generate more useful drills and foster the relationship between the two. The chief officer should be left to conduct the drills, because this will show your trust; in turn, the chief officer is more likely to open up and report difficulties. Gaining the chief officer's confidence and trust will avoid the risk of the Master discovering problems only when something goes wrong, which could be the difference between a disaster and a near-miss.

Relationship with the crew

Smaller crews make it vital that teamwork is efficient and fluid. The need to build strong relationships with heads of departments should not hide the importance for the Master to remain connected with the other members of the crew. The challenge is to strike the right balance between staying in touch and becoming too friendly or interfering.

The relationship with the crew needs to be exercised with utmost diplomacy and sensitivity. The Master needs to be a good communicator. For certain aspects you will deal with the crew directly, but others must be left to the heads of departments.

Direct relationship concerning duties

For matters such as bond store issues and the crew effects list, the Master exercises direct authority and so has the chance to be in direct contact with the crew. This is an opportunity for the crew to bring up grievances which you might otherwise be unaware of. It is good policy to ask crew members concerned if they have already spoken to their head of department. You can then decide whether to tackle the issue yourself or refer it back to the head of department. On these occasions, the Master must think carefully before speaking.

Heads of departments

In communicating with crew members, the Master must take care not to alienate the heads of departments. If you have established good relationships with senior officers, this will avoid mishaps of this kind. You need to be sensitive to the needs of the crew and should listen to them attentively, but then pass the matter to your direct subordinates for further action.

It is said that a tight ship is a happy ship. Such a ship is where everybody knows where they stand and knows what is expected of them. It is up to you as Master to make sure this is true of your ship. You have to be a good communicator to establish good relationships on your ship.

Among your many functions, you as Master must be a leader, listener, confidant, mentor, diplomat and arbitrator.

MLC pastoral care

By Reverend Canon Ken Peters

A primary responsibility of the Master in command is the health and safety of the crew. With the authority over their living and working conditions comes the responsibility of ensuring that the crew are treated in line with best management practices and in compliance with the law.

For centuries management of crew was largely left to custom and practice. Now the MLC sets out clear standards, regulations and guidelines for the well-being of crew. Other articles in this section set out in more detail elements of the MLC that companies need to follow.

MLC is a comprehensive package that not only consolidates previous ILO conventions but also introduces new legislation to ensure that it addresses current issues. It covers minimum requirements for seafarers to work on a ship; conditions of employment;

accommodation; recreational facilities and catering; health protection; medical care; welfare and social security and compliance and enforcement. It places a clear set of obligations on shipowners and sets down the responsibilities of flag and port states.

Masters must first ensure that all seafarers aboard the ship are more than 16 years old, have valid medical certificates and are properly trained and qualified for the job they are doing. For those seafarers under 18 there are strict conditions attached to the type of work they undertake. Every seafarer must have a valid, legal and fair employment agreement in accordance with the conditions laid out in Standard A2.1.

The seafarer employment agreement (SEA) covers wages, hours of work and rest, entitlement to leave, repatriation and other essentials. It serves as the reference document should there be any dispute over the terms and conditions of employment. Taking care to verify that the SEA conforms to acceptable standards should reduce the risk of port state control (PSC) or unions disrupting the ship's routine and business continuity. Successful verification will give the Master confidence that the crew is content with the contractual rights and obligations set down in the SEA. However, in the event of disputes it provides seafarers with an accepted complaints procedure.

Maritime welfare organisations point to irregularity in wage payments as the single most contentious issue disrupting the relationship between shipowner, Master and crew. Payment of wages in conformity with the SEA is essential to the well-being of seafarers. Late payment, part payment or any other irregularity is guaranteed to cause discontent among seafarers. Crew members' efficiency will be affected if they are worrying about the financial security of their families at home. Keeping the crew content is best achieved by ensuring payment is on time and in full. If there is any administrative difficulty, either within the company or through the banking system, the crew should be notified immediately. It is vital to maintain the trust of the crew and not to allow the integrity of the company to be called into question, as this will undermine the relationship between Master and crew.

Some of MLC's most contentious provisions concern hours of work and rest (Reg. 2.3), as these directly affect the ship's working routine. Matching the complexities of watch rotations to finite human resources while abiding by the MLC and STCW rules can prove extremely challenging for Masters.

Maintaining safe navigation lies at the heart of these provisions together with the avoidance of fatigue – a contributing factor to many maritime casualties. Overwork and interrupted rest periods seriously undermine seafarers' ability to perform their duties effectively. However, the person most likely to be affected by fatigue is the Master, who often feels the need to be present on the bridge for many hours. This is especially true during short sea passages when calling at several ports in quick succession. Masters must be aware of the dangerous effects of fatigue when organising the ship's routine.

Another difficult provision to organise is crew leave. Reg. 2.4.2 states: Seafarers shall be granted shore leave to benefit their health and well-being. In an era of minimal manning and fast turnaround, this can be administratively demanding, as the Master has to balance this requirement with the necessity of maintaining the ship's operational capacity. Indeed, the same regulation notes that shore leave for seafarers has to be consistent with the operational requirements of their positions. Few things make seafarers so disgruntled as denial of shore leave.

Repatriation is another sensitive welfare issue. Denial of repatriation can inflict a heavy psychological blow to a seafarer. Masters will appear uncaring if unwilling to consider requests for early repatriation. Apart from the termination or expiry of the SEA, legitimate grounds for such a request are, in Standard A2.5 (c) when the seafarers are no longer able to carry out their duties under the employment agreement or cannot be expected to carry them out in the specific circumstances. You must use your professional judgement when considering whether to grant repatriation. Once again, a balance must be struck between the ship's operational requirements and the seafarer's pastoral needs. Denial of timely repatriation of crew at the end of the contracted period because of delayed relief would be considered unacceptable to most and could damage an operator's reputation. The longer the delay, the more likely this will trigger a complaint, either through the onboard complaints procedure (Reg.5.1.5) or the onshore complaints handling procedure (Reg. 5.2.2). The latter is more time-consuming and can cause significant disruption and delay.

For many seafarers the most important person aboard is the ship's cook. MLC Reg. 3.2, plus the relevant standard and guidelines, concentrates on diet. This aspect of shipboard life needs careful attention, as health, well-being, fitness and mental alertness are all dependent upon a healthy balanced diet of appropriate quality, nutritional value and quantity. As well as the quality, preparation and cooking of the food, it is important that the menu reflects the crew's cultural and religious backgrounds (Reg. 3.2).

Additionally, the provisions of MLC embrace culture and religion more generally. Being knowledgeable about the customs and practices of the cultural groups and faith communities represented on board, and being sensitive to their needs, will improve professional relationships and enhance the esteem in which the Master is held.

Health protection, medical care, welfare and social security provisions are all welfare matters that directly affect the crew's well-being. To ensure the medical health of the crew, Masters need to be fully aware of Standard A4.1. Member states of the ILO have to ensure that seafarers working onboard a ship have access to shore-based facilities and services to secure their health and well-being. Once again, shore leave is of considerable importance in easing the isolation of serving at sea. Masters have access to the various maritime missions that can help seafarers with their welfare needs. The Apostleship of the Sea and the Mission to Seafarers are probably the two best known global networks of seafarers' centres and port chaplains.

The final section of MLC 2006 concerns compliance and enforcement. Failure to comply with the convention could lead to a disruptive and costly detention by port state control. It is worth remembering that compliance means that the living and working conditions of the crew are of the standard that you as Master expect for yourself.

Many of the authors of the MLC believe that a well-trained and happy crew results in a safer ship.

Health, well-being and hygiene
By Dr Toby Abaya

Health and wellness take on new meanings in the seafarer's work environment as the challenges, the responses to them and their outcomes will differ from those of most land-based occupations.

There are three components to health and well-being: structural (anatomy), physiological (nutrition and biochemical) and psychological (mental health/emotional). This triad can be interrelated, and work together to achieve total wellness.

The pre-employment medical exam (PEME)

All seafarers have to undergo this examination before they go on board. It is important to recognise that a PEME does not involve the usual doctor-patient relationship. Generally, people go to a doctor when they are unwell, and they tell the doctor about their problem. By contrast, the seafarer may view the PEME as an unwelcome necessity that must be undergone if they are to be hired. They may be reluctant to reveal their complete medical or surgical history, fearing that declaring a past or current medical condition may lead to rejection or an unfit status. In many cases, seafarers may try to hide medications or deny they have an illness.

With a complete history doctors are better able to manage the patient. Doctors can advise seafarers about possible issues they may face and recommend diet, exercise and lifestyle modifications so as to maintain good health.

Diet

Diet is part of the triad because it supplies the nutrients that keep one's biochemical, structural and psychological state in balance. Breakfast is considered the most important meal of the day, providing up to 25% of total energy requirements for the next 24 hours and ensuring concentration and dexterity during the morning. Carbohydrate sources such as bread, pasta and rice are important and whole grain options should be chosen, if possible. Skimmed or half-fat milk is a great source of protein for maintaining muscular and skeletal strength.

Watchkeeping is an important part of the work of the seagoing professional. Alertness and vigilance are essential and diet can play an important role during this crucial assignment. For preference, at the beginning of the shift eat protein-rich food such as skimmed and half-fat milk, yoghurt and lean meat. These give a slow, steady release of energy. Later on in the shift eat carbohydrates such as bread, potatoes, salad, pasta salad, fruit and whole grains. These encourage better sleep. Spicy, fried and fatty foods should be avoided. Caffeine (found in coffee, tea, soft drinks and energy drinks) can help keep one awake, but more than 400mg a day – equivalent to three cups of coffee/tea, a litre of soda or two energy drinks – can cause insomnia, palpitations, restlessness and trembling.

Diet affects weight, and being overweight increases the risk of cardiovascular disease and diabetes. However, dieting alone will not lead to weight loss. The combination of a healthy diet and regular exercise is the best way to maintain a good weight.

Hygiene

Personal hygiene is taking care of one's body, keeping it clean and minimising the spread of disease. Regular handwashing is essential. Your hands can spread many contagious diseases, so it is very important that everyone onboard washes their hands before eating, after regular work and as often as possible when exposed to dirty and unclean surfaces.

In places where infectious diseases are common, personal hygiene becomes especially important. Ships travel to many parts of the world where cleanliness, hygiene and medical care are poor and where there is a risk that infectious diseases may be brought on board. Should you suspect this has happened, isolate the crew member concerned and obtain advice immediately from shore-based medical experts.

Dental calls account for more than 10% of all referrals on board. This is an alarming rate since dental disease is almost always preventable. Almost 90% of referrals for tooth and gum disease stem from poor dental hygiene. Reminders to crew to brush their teeth at least twice a day should minimise the need for dental calls.

Tattoos are an old tradition among seafarers. New technologies have made it safer, but not all tattoo artists employ safe techniques or clean instruments and materials. There is the risk of transmission of blood-borne diseases such as hepatitis B and HIV if the tattooist uses the same needle on more than one customer. Cleanliness should be a guide when choosing a parlour.

Mental health

Although few seafarers need to be repatriated because of mental issues, Masters should be aware of the signs and symptoms of impending or continuing psychiatric problems. They include insomnia, uncommunicativeness, lack of appetite and rapid or dramatic shifts in feelings or mood swings. Masters should talk to anyone presenting with these signs and reassure them. The internet (if available) can be a valuable tool, as sometimes just communicating with family and loved ones may help. Early intervention is the key.

Shore offices may be able to say whether the person concerned has any history of mental health issues. Seafarers should also be asked (carefully) if they are taking any medicines that you as Master have not been told about. As already mentioned, historical medical issues are often not revealed for fear of an unfit status.

Special care is needed when handling possible mental health issues. People can be very vulnerable when depressed or anxious. The source of that anxiety may be something as simple as a lack of communication from their family.

Hours of work and rest

These should be closely monitored, not only as a statutory requirement but because many people at sea tend to want to 'get the job done' and at times can work excessive hours to their own detriment. Any crew member who is feeling less than 100% should be monitored in regard to their working hours to ensure that they do not overdo it and end up in the sick bay. See pages 44-47.

Medical emergencies

Satellite communications and the internet have transformed communications at sea, and telemedicine has proved to be of great value in managing medical emergencies. The medical staff will need to know the time frame of the incident/illness and the vital signs of the patient (heart rate, respiratory rate, blood pressure, temperature). If pain is present, give an indication of intensity (usually 0-10). Be as exact as you can when identifying where the medical problem is on the patient's body. To help with this, the abdomen can be divided into quadrants and the upper and lower extremities can be further subdivided. Working methodically can save time and help the medical team assist the ship in the most efficient way.

You are responsible for the ensuring the health and well-being of your crew during your command and there are several articles in this section of the book to help you carry out this important duty.

Exercise

Encourage your crew to exercise – and lead by example. Acquaint yourself with what is possible onboard: circuit training is a great no-equipment workout for seafarers. Body weight movements optimise functionality, which means that over time, seafarers will become more nimble, strong and healthy.

It is possible to formulate a workout with a variety of exercises to engage all major muscle groups, which is easy to perform on a vessel despite the physical restrictions.

One example:

1. Begin with a warm up; jumping jacks or running on the spot
2. Lunges
3. Push-ups
4. Mountain climbers
5. Squats
6. Burpees
7. Leg raises
8. Plank

Spend 45 seconds on each exercise, with a 15 second break in between.

Section 4 | 87
Day to day management – personnel

EXERCISING ON BOARD

Spend **45** seconds on each exercise, with a **15** second break in between.

BEGIN WITH A WARM UP

1. RUNNING ON THE SPOT
2. JUMPING JACKS
3. LUNGES
4. PUSH UPS

THE NAUTICAL INSTITUTE ON COMMAND

Section 4
The Nautical Institute on Command

The International Seafarers' Welfare and Assistance Network (ISWAN) runs a Training on Board project see www.trainingonboard.org

More ideas can be found on the human element project Alert! website www.he-alert.org which is a campaign by The Nautical Institute, sponsored by the Lloyd's Register Foundation.

Many P&I clubs publish useful information on exercise – explore their websites.

Food and catering

By Tapan Kumar

Napoleon Bonaparte famously said "an army marches on its stomach". It is no different for seafarers. The performance of a ship hinges on its crew's morale, physical health and emotional well-being, which in turn depend on seafarers receiving good quality nutrition on board.

Few Masters are briefed on the value of good catering management. Nevertheless, you have a crucial role to play in motivating the chief cook and catering crew in providing safe, healthy, tasty meals within the allocated budget.

The Maritime Labour Convention (MLC) 2006 instructs flags to establish national standards for the quantity, quality, nutrition and variety of food on board, taking into account religious and cultural diet preferences. The convention requires flags to ensure that cooks and catering staff are properly trained and qualified. MLC also requires documentary evidence of the Master's weekly inspection of food handling spaces for hygiene and food safety.

The weekly Master's inspection of accommodation, galley, refrigerator room, provisions stores and drinking water source constitutes the first step to food safety. The Master must ensure that:

- Galley floors, exhaust filters and gutters, tables, equipment, cutlery and crockery are spotless and free of grease, dust and food stains
- Food is kept in tightly closed containers. Food waste should be in closed non-combustible metal bins
- Galley staff are clean-shaven, short-haired, scrubbed, with nails trimmed and hands clean and free of injuries and calluses. Uniforms should be scrupulously clean
- Refrigerator rooms are maintained at a maximum temperature of -18°C for fish rooms, -12°C for meat rooms and +4°C for vegetable rooms. Any frosting on the refrigerator room door packing and on evaporator coils should be removed
- Galley staff undertake a thorough monthly cleaning, de-infestation and squaring up of all food stores. Inventory taking should be part of the routine. Moisture and infestation are health hazards in vegetable and dry provision stores. Moisture promotes fungal growth, and uncovered food encourages infestation by small insects, cockroaches or rodents

- Expiry and use-by dates are monitored with the first in, first out (FIFO) rule applied
- Wastage is minimised. Large quantities of galley waste can cause odours, infestation and health hazards
- Food is prepared and served fresh. Cooked meals should be served hot, never reheated unless it is absolutely essential. In that case chill food and refrigerate; then reheat it to its original cooking temperature before serving.

Meal times are sacrosanct. However, changes in normal working routines, particularly while in port, need to be taken into account. The Master should discuss such changes with the chief cook to ensure that everybody is fed.

The majority of the world's seafarers originate from eight broad dietary regions: North America and UK; Western Europe; Eastern Europe and Mediterranean; Middle East; Indian sub-continent; Philippines; Thailand; Burma, China, Korea and Japan. Each region demands a separate stock of victuals and its own menu. Catering staff should understand the nuances of the crew members' own diets.

The following principles of nutrition are universal:

- A good balance of proteins, vitamins, minerals and carbohydrates
- Fresh fruit (not canned or juiced) and salads for dietary fibre
- Avoid greasy, fried, reheated or processed food and butter as sources of unhealthy ('bad') cholesterol
- Processed foods, such as noodles, packet soups, preserved meats and fish, sauces, processed cereals, cookies and biscuits, and jams are tasty but processing causes these foods to lose much of their nutritional value
- Meat, fish, eggs, milk, lentils and cheese provide essential protein. White meat, such as chicken, fish and seafood, is low in bad cholesterol. Red meat, such as beef, mutton and pork, is high in saturated fats, calories and iron, and therefore high in bad cholesterol
- Vegetables provide vitamins and minerals. Stir-fried or partly boiled/cooked vegetables retain more nutrients than deep-fried vegetables
- Whole-wheat and whole-grain staples retain more nutrients and provide more fibre than sticky glutinous rice, white bread or finely ground wheat products.

Crews are served three meals with two coffee breaks. To prevent hyperacidity, food – such as fruit, sandwiches, salads, milk, cheese and juice – should be provided in the fridge for the night watchkeepers.

Equatorial and tropical climates tend to be humid and hot, especially in summer. In the tropics, seafarers can get quite dehydrated working on deck or in the engine room. They tend to prefer foods with more water and mineral content such as greens, above-ground vegetables, fruit and fluids.

In cold, temperate zones, the body needs more protein and carbohydrates. The crew will instinctively prefer more meat, seafood, milk, cheese and root vegetables to provide the body with strength to combat loss of body heat and energy.

On most ships, water is produced by a desalination plant called a freshwater generator. This water is distilled and free of minerals, so for drinking purposes it needs to be dosed with mineral tablets or an online mineraliser. If this is not available, bottled mineral water, which contains minute quantities of essential minerals, can be used as a supplement.

Taste, freshness of preparation, presentation and variety are essential. The Master can gain useful feedback through a monthly crew survey that measures these four parameters on a scale of 1 to 5. You should employ tactful encouragement and positive reinforcement to get the best out of the cook. Direct interaction between crew and cook is best avoided – food can be a very sensitive matter.

Sourcing good provisions is a daunting administrative challenge and needs good planning. At least two independent quotations should be obtained from ship chandlers.

Recommended replenishment frequencies are:

- Dry provisions (including juice, sauces and long-life milk) – 2 months
- Fresh vegetables and fruit – 15 days, if voyage length permits
- Meat, fish and poultry – 1 month
- Frozen food – 2 months.

It is important to prevent wastage and spoilage. Long-standing stock should be consumed first, supplemented with frozen stock on the FIFO principle. Physical stocktaking is crucial for optimising budgets, using a simple Excel sheet for monthly inventory, consumption and budget reports. Budgets should take into account dietary habits, trading regions and cultures. An example of best practice can be found at www.nautinst.org/command

Prices can vary widely from port to port, but little can be done about this beyond buying lower quantities at expensive ports and deferring bulk purchases to cheaper ones. The rolling consumption rate tends to even out over a four-month period. It is, however, counter-productive to the ship's performance to cut back on a healthy, balanced diet simply to save money. The aim of budgeting should be efficiency, not lower consumption.

Onboard training and development

By Captain André L Le Goubin

As Master, it is essential that you have the necessary practical knowledge and training to safely command the vessel. Now, you must also take responsibility for the onboard training and development of those under your command. This article will briefly discuss how we can do this onboard today's modern merchant navy vessels.

You may know the terms 'experiential knowledge' (knowledge gained from experience and reflected upon) and 'mentoring' (the transfer of experiential knowledge without designated reward) but may not understand how they are relevant to you as a new or aspiring Master. Look back over your career so far and consider how the knowledge you have gained from your experiences onboard and taken the time to reflect on, has given you the confidence to put into practice all you learned at college, from books or online, and by simulation. Think also about the people who have mentored you and the experiential knowledge you gained from them.

As a new Master, you are going to gain a lot more knowledge and very quickly! Every manoeuvre you undertake, every meeting you have and every conflict you resolve (just to give a few examples) should give you knowledge. You need to take a few minutes out of your busy schedule to reflect on how it went, what you learnt from it and, if necessary, how you would do things differently next time. If things didn't go too well, gain experiential knowledge from it and then move on. Experiences can be good or bad, but the knowledge that comes from experience can only ever be good.

Take time to reflect on the essential knowledge you think you need as Master but lack. For example, a number of new Masters have experienced accidents when approaching an anchorage or pilot station when vessels could not be slowed down in time. Do you know how quickly and easily your vessel slows down? To find out, I suggest you arrive at your next pilot station 30 minutes ahead of your given ETA and practise slowing your vessel down. Take time to see if the way comes off easily and the reaction when you put the engines astern. Your next arrival will be so much easier and you can pass that experiential knowledge on to others, such as the pilot. This is particularly important if your vessel is difficult to stop.

Through research into accidents and incidents, I have identified four areas where it is critical that a new Master has the necessary knowledge, practical training and experience from the day of taking up command. They are:

- Anchoring the vessel
- Manoeuvring to embark a pilot
- Manoeuvring the vessel in a channel or fairway
- Handling the vessel in heavy weather.

The best way to gain the necessary knowledge and experience is by undertaking these tasks as often as possible before taking command. Some companies allow time for this where the aspiring Master is taken out of the ship's complement and sails as a trainee Master for a number of months. This is good but the training period is rarely, if ever, long enough. If you are already in command and feel you need additional training, ask your company to send you on a shiphandling course and investigate what practical books are available on board or through the company office. Pilots are often willing to share their knowledge when approached. Most of all, after every manoeuvre take a few moments to reflect on how it went and gain the experiential knowledge.

Even if you delegate day to day training responsibilities to one or more of your officers, as Master you have the ultimate responsibility for the training and development of those under your command. A knowledge-sharing system onboard, such as mentoring, will enable each officer to undertake tasks in preparation for promotion. Once promoted and settled in the new rank, I believe all officers should begin training for the next promotion and the newly promoted officers should begin training their successors.

With a bit of planning, much of the onboard training required can be incorporated into the daily routine and will not require extra time to complete. Passing on knowledge in this way should never take more than 10 minutes extra (unless you want it to), which is the time it takes to smoke a cigarette or drink a cup of coffee. When you, or one of your senior officers, undertake a task try to have a junior member of staff to assist so they can learn. Better still, if you feel confident, let them undertake the task under your (or your senior officer's) supervision. By integrating mentoring into the daily working of the vessel, your staff will adopt this as routine rather than something out of the ordinary.

Some examples:

- Take 10 minutes to discuss your standing orders with each of the watchkeeping officers
- Spend a few minutes each day with each of the watchkeeping officers. Perhaps take a cup of tea or coffee on the bridge rather than in your cabin. Ten minutes a day is enough for them to get to know you and to be sure of your reaction when they call you.
- Ask the chief officer to supervise the second or third officer letting go the anchor. When they are competent, the chief officer could understudy you on the bridge and bring the vessel to anchor under your supervision.
- Supervise the chief officer picking up the pilot.
- Ask the second officer to guide the third officer in preparing a passage plan.
- Ask the chief officer to guide the second officer in preparing a cargo load (or discharge) plan.

Use of a common language is really important for the training and development of everyone onboard now that a substantial proportion of merchant vessels are manned by multinational staff. As Master, insist on the use of a common language whenever possible and you will benefit in many ways. Effective communication is vital to the smooth running of the vessel but can be so difficult unless you can establish and practise the use of a common language at every opportunity, especially socially.

It is well recognised, although not always acknowledged, how busy the Master is. If you are already in command I hope that you will take some time out to reflect on your own training needs and those of your crew. If you have not yet taken command, decide what practical training you require and take advantage of every opportunity onboard to gain that knowledge. By addressing training needs within the daily operation of the vessel and taking every opportunity to gain and pass on experiential knowledge it will be possible for you, and those under your command, to gain the necessary experiential knowledge. I hope that you will encourage those serving with you, in whatever rank, to do the same as part of their continuing professional development (CPD).

For more information on mentoring, see *Mentoring at Sea: the 10 minute challenge* by André L Le Goubin published by The Nautical Institute.

For more information on The Nautical Institute's CPD scheme visit www.nautinst.org/CPD

Discrimination

By Captain Wendy Maughan

Masters need to be alert to the potential for discrimination among the many people of different religions, races, sexual orientation and age they will encounter onboard ships. You must be aware at all times of the backgrounds of those onboard, the possibility of differences of opinion and conflict and have strategies to deal with these.

The company should have guidelines and policies to deal with all forms of discrimination: make sure you are aware of these. If you do not have them, seek help from your employer. Should you become aware of any form of discrimination on your ship, it is you as Master who must be prepared to deal with incidents promptly. If left to fester, incidents will only get worse and could lead to tragic consequences as has happened in the past. You must always be vigilant. See pages 101-104 for more detail on dealing with conflict.

All types of discrimination are extremely hurtful to the victims and disruptive to the operation of the ship. You will need to handle all incidents sensitively and they will all require careful thought and consideration. Remember that you must be, and be seen to be, unbiased in all personnel matters, however difficult they may be for you to handle. If you need help or guidance, contact your employer (or the employer of the affected seafarer) or your technical manager.

As a serving Master, I hope the following case study in sex discrimination will help put some of this in perspective. From the standpoint of a female having spent a life at sea and working ashore in ports, I have to admit that sex discrimination does still exist. Over the years though it has diminished and as I have become older (if a woman ever does age!!) people seem less inclined to question one's abilities!

So how did I find things when I first went to sea? How did I manage the discrimination? And how do females in the shipping industry move forward?

In 1978 when I first went to sea the UK's Discrimination Act had only just been passed and it had an immediate and major effect as it enabled females to join the merchant navy for the first time. Of course not all sex discrimination stopped overnight and there were hurdles along the way to cadetship, getting a job and gaining promotion. Many women are still battling to get that first job, especially where there is no legislation to help them.

So how can one level the playing field when dealing with actual or perceived sex discrimination? To get upset every time it happens would make one bitter. To take legal action would result in spending more time in court rather than doing the job!

My advice, or at least what's worked for me, is to:

- Be aware that selection on the grounds of sex, for whatever reason (some valid, some not), does still exist, so you need to be proactive to compensate

- Without being pushy, persevere with what you're aiming to achieve (make the effort to phone, email or visit the decision makers)
- Be patient – don't expect to get what you want straight away, you may need come back later when 'attitudes' (or sometimes even people) have changed
- Find ways to turn a perceived disadvantage into a positive incentive
- Don't waste your time and effort worrying about what you can do to minimise the problem – it's their problem, not yours, let them sort it out!

Having landed the job of a lifetime the next hurdle is to keep it!

No matter who you are, or whether you are male or female, you want to prove your worth. On the other hand you're new to the position and no matter how experienced you are, each organisation and team works differently. As a female the difference is that all eyes will be on you analysing your every move.

Some colleagues will see you as fair game, especially if they consider themselves to be God's gift to women. When I was younger I found these guys the hardest to deal with. They will be delighted that you are joining the team and will be the first to greet you and offer their unlimited support. Be aware they will want something in return. One man would not believe that I was onboard to work – every morning he would make his advances. A rebut, followed by less than feminine language landed on deaf ears – he was not taking no for an answer. In the end I enlisted help from other colleagues in my quest to make the position clear. Matter closed, one would think, but I can assure you he spent the rest of the trip finding fault with my every move!

Others may not agree with women in a man's world and when you arrive they may avoid contact; if you handle it correctly these guys may be your strongest ally. Most importantly, they could support other females following in your footsteps. They will be slowly won over by observing that you just get on with your job and always do it to the best of your ability – and I should know, I'm married to one!

When the job is going smoothly, relationships tend to settle down and good working arrangements develop. However, when tensions mount due to operational problems, opportunities of promotion or financial restraints, colleagues will start looking for a scapegoat to vent their frustrations. I have found this negative energy will often be directed towards a female member of the team. This is a difficult situation to deal with and often time and understanding are needed to resolve these tensions.

Of course these are the extremes and your relationship with your colleagues is paramount for your success in the industry. Most importantly, however, because females are a minority player, your success in the industry is paramount to the acceptance of females by the industry in the future.

So how can we measure the acceptance of females in the shipping Industry?

I believe general acceptance has improved dramatically. Twenty-five years ago I was driving a hydrofoil between Southampton and Cowes and passengers had to pass the hydrofoil commander, ie me, when going to their seats. I frequently received a double take, which in rough conditions, was followed by an expression that indicated: "Should I catch the next ferry? Should I put my life in the hands of a female driver?" Having paid for their tickets they at least didn't get off! In contrast, when I recently hosted a school visit to the bridge of my ferry the kids were completely unconcerned that I was a female. It's amazing what one generation has achieved and, although females in the shipping industry are still a significant minority, it bodes well for the future and is heartening to see.

Lastly I would like to ask all females within the industry to consider the importance of their actions. I will always remember the damage done to our interests by a female I sailed with who used her feminine wiles to achieve her aims. Despite the progress already made females are still being watched and it is imperative that my female colleagues continue to present a professional image and do the job to the best of their ability. Please do not lose your femininity but aspire to be the best role model that you can be. We are pioneers in our profession and have a responsibility to those females who wish to follow in our footsteps.

Appraisals

By Captain Sriram Rajagopal

Performance appraisals are among the most potent tools in the hands of Masters and officers, to help to improve the performance of other seafarers you work with. However, the effectiveness of appraisals depends to a large extent on the way they are carried out. Here I share some lessons that I have learned onboard while conducting appraisals – as well as being at the receiving end.

It is common in shipping companies today to have a system of open appraisals, rather than the confidential reports used in the past. The reason for this change is simple – confidential reports only told the company a senior officer's opinion of a particular seafarer's performance onboard. They do not offer seafarers an opportunity to improve or even give any information to the seafarer. In the long run they are not very productive and do not help to improve people.

Realising this limitation, most companies have migrated to an open system of appraisal, where seafarers sit with Masters or superior officers, who conduct appraisals about their performance onboard.

It is very important that there is consistency in these appraisals so, when seafarers move from vessel to vessel or are appraised by different Masters or superior officers, candidates recognise the process and understand what is expected of them.

People being appraised can be nervous and it is essential that those conducting the appraisal are calm and professional. It may be that as a new Master you have not conducted an appraisal before but a senior officer onboard has. In this case, don't be afraid to ask for advice and help to be mentored through the process. It may be sensible in these circumstances to have two making the appraisal, the new Master and the more experienced senior officer.

Some simple tips to remember:

- Have a single form throughout the company – check the ship's stock isn't out of date
- The appraisal is two ways – ask for comments and receive them with good grace even if you are criticised
- Evaluate positive traits as well as negative aspects, emphasising areas for improvement
- Inform seafarers of strengths so these are developed
- Examine weaknesses and allow opportunities for improvement
- Brief the candidate, explaining the aims of the appraisal, so they feel comfortable with the process. The result of the appraisal depends, to a large extent, on the manner in which the candidate takes it
- Be prepared. Make notes beforehand containing points you would like to discuss. Use bullet points and the company format as a guide.

Start with a good word. Everyone, even the worst worker, has some good traits. Discover them and start your appraisal by referring to them. Your appraisal will have a better effect if the appraised seafarer feels that you are being fair and unbiased. This is also a helpful exercise for you; it enables you to see merits in each person and develops a non-judgmental demeanour – an important trait in Masters.

Be specific and stick to the seafarer's work – not general behaviour or nature. Avoid making vague comments such as: "You are lazy." Instead concentrate on specific aspects of the seafarer's work. Give instances where work was not right, with an explanation of what was expected of them. This approach also gives seafarers an opportunity to identify where and how they can improve.

Avoid being too negative. This is a pitfall that many officers fall into by appraising on just one negative trait. Remember, the appraisal is about seafarers' work, and any traits that specifically affect their performance onboard. Some traits are useful to have as a seafarer, and some may be undesirable, but avoid making them the centre of the appraisal.

For example, a seafarer has been found to be often rude or brusque in their manner of speaking with seniors. Explain this, citing specific instances, acknowledge merits – they may be rude, but hardworking and knowledgeable. Do not let their negative traits make you biased towards their performance.

Let it be a dialogue. Remember appraisals are not meant to be a one-way conversation. To be truly useful, they should be a dialogue, a discussion where the person being appraised is given a fair chance to explain their view of how things have gone over the previous year. Involve seafarers in finding solutions to shortcomings. Ask if they need any assistance or training. You will be surprised how forthcoming they might be.

Avoid being too positive. Sometimes individuals may be appraised completely based on a perceived positive quality, overlooking the employee's shortcomings. This is wrong and must be avoided. For example: seafarers with positive dispositions but poor work records may get positive appraisals simply due to their cheerfulness.

Appraisals are not meant to be your complaint board. Avoid launching into a tirade about all that is negative about the seafarer. No one likes to constantly hear complaints about themselves. Instead, see how your criticism and comments can be constructive. Always give a way out to the seafarer. This does not mean you don't explain lapses, but keep it professional. Otherwise, they may choose not to respond or the appraisal might turn into a two-way verbal fight! You do not want either of these.

Remember that at the start of the appraisal, it is your opinion of the seafarer that will dominate the conversation but that there are two sides to any story and the purpose of appraisals is for continuous improvement. Don't take criticism personally. Be professional. Listen to the seafarer's side, and try to improve the situation.

Here are some pitfalls that might result in bias in appraisals:

- Appraising purely on recent events
- Personal bias, either positive or negative, based on favouritism or dislike
- Similarity bias if the seafarer and officer are alike.

Do try these strategies when you conduct your next set of appraisals. Remember that when used properly, appraisals are a helpful method to improve performance onboard. Keep improvising and improving the performance of your fellow seafarers – it will help you in the long run.

Discipline

By Iain Macleod

An effective disciplinary process is an essential management tool for any organisation. Problems with conduct or performance will arise and these must be resolved early to ensure the continuing effectiveness of the organisation. The disciplinary process should be a tool for the daily management of the vessel, not a weapon of last resort. An effective disciplinary procedure will not be administered only by the Master; anyone involved needs to be aware of the procedure and be confident in its use.

In cases of serious misconduct or disputes, the process may escalate beyond the vessel. The seafarer's trade union may become involved. The Maritime Labour Convention 2006 gives the seafarer the right to contact port state, flag state, or trade union at any stage of a dispute. A litigious culture exists in some countries; this may be reflected in the position adopted by a maritime authority.

The most effective approach to dealing with disciplinary matters may be summarised in three stages of action: avoid, prepare and conduct.

Avoid

Early intervention is the best way to prevent poor performance or conduct becoming a serious issue. It is essential that all officers and petty officers are engaged with the process. Poor conduct or poor performance should be addressed at the first time of occurrence. Some key points:

- Speak with the seafarer as soon as possible after it occurs
- State clearly what is unacceptable; explain the required standard; check that the seafarer understands and knows how to achieve it
- Acknowledge good performance. Communication should not be only criticism; praise reinforces the expected standards of performance or conduct
- In a dispute between seafarers, make sure both parties are spoken to and that you understand both sides
- Food can be a source of problems; provision of inappropriate food is insensitive and may be perceived as lack of respect
- In some cultures, respect for elders and seniors is essential and deeply ingrained. Some seafarers may be reluctant to speak up to deliver bad news or unwelcome information. If the senior appears to ignore or discount such input, the seafarer may not speak out on a future occasion when required. Cross-cultural training is available; much useful information is available on the web. Aspiring senior officers should familiarise themselves with the principles
- Disputes often arise as a result of cultural misunderstandings. These can be the result of gestures, such as finger pointing, reference to race or gender and may even be in the form of address
- Language may become a barrier. Avoid feelings of exclusion by insisting that all conversations are conducted in the common language of the vessel.

Prepare

Not every issue can be avoided and disciplinary issues will arise. We all wish to be treated with respect and to be valued as human beings.

- When a mistake is made, explain failures in private and give guidance on how to improve
- If there is a disciplinary procedure, follow it
- Some issues may be affected by requirements of flag, contract, union, MLC 2006, or the company, so the Master should communicate with company at an early stage
- In the absence of company disciplinary procedure, I recommend use of the UK's Merchant Navy Code of Conduct, revised in August 2013. Prepared and approved by owners' associations and trade unions, it is a robust and fair process which will withstand examination in any court. It may be freely downloaded from the web.

Certain features should be present in any good disciplinary process.

- Issues should be settled at the lowest level, informally where possible
- Keep written records of each incident. This provides the foundation for escalation, should the problem recur
- Investigate the matter thoroughly before any sanction is imposed
- Allow seafarers to give their own account and to question others about their version

- A range of sanctions should be available depending on the severity of offence
- Except in the most serious cases, where the safety of the vessel or the well-being of the crew may be affected, an offender should be given an opportunity to improve
- Certain offences may be defined as gross misconduct and warrant dismissal from the ship. These may include violence, theft or violation of drug and alcohol policy. Procedure must still be followed
- Where the sanction is dismissal from the ship, the company must be informed. The Master may dismiss from the ship, but only the employer may dismiss from employment and will have to conduct a separate investigation and disciplinary process with the employee ashore.

Conduct

The procedure for conducting both the investigation and the disciplinary hearing is important. Follow your written procedure.

- Set aside time, hold the meeting in private and avoid interruptions. Have another officer assist by taking notes
- Ensure the seafarer is allowed to have a colleague present. If this is declined, record that
- Explain the nature of the offence clearly. Give the seafarer the opportunity to provide their own account. Either the seafarer or the Master may call other witnesses and the seafarer should be given the opportunity to question their accounts
- Give the seafarer the opportunity to state their case
- Check continually throughout the process that the seafarer has understood what has been said. When all evidence has been presented, the Master should ask the seafarer to confirm that ample opportunity has been given for stating his case
- Adjourn the investigation meeting to consider the evidence and your conclusions. Record all statements and points of discussion and give a copy of the minutes of the meeting to the seafarer
- At the reconvened disciplinary meeting, announce and explain your decision. Remember that, in employment matters, you are allowed to make a judgement based on the balance of probabilities; it is not necessary that a case be proved beyond all reasonable doubt
- Explain that you have considered the evidence presented, the severity of the issue and the past record of the seafarer and that you have considered various possible sanctions before choosing this particular one
- Provide a copy of the minutes to the seafarer
- Retain all notes and copies of statements and enter a record of the proceedings in the log
- Send copies of all notes and materials to the company, for entry into the seafarer's employment record and to facilitate any further actions which may be required by the company, as in the case of a dismissal from the ship

- If the sanction is dismissal from the ship, consideration should be given to delaying the disciplinary meeting until such time as the seafarer may immediately be put ashore, rather than having someone on board knowing they have been dismissed
- Do not shirk the decision to dismiss from the ship. You have a duty to the vessel, to your crew, your colleagues and to the company not to tolerate sub-standard behaviour or performance.

In any situation where there is subsequent involvement of outside parties, our best defence is to be able to demonstrate that we have followed a robust and fair process in reaching our decisions.

Managing conflict at sea

By Alison Williams

Positive conflict in the workplace can be a good thing. Workplace conflict is unavoidable – stifled conflict leads to greater conflict, while the elimination of any conflict suffocates innovation. The key to success is to ensure that you recognise when conflict is happening, understand why it is happening and deal with it effectively. A solution can usually be found through open discussion, but sometimes you may need to enlist the help of others to ensure a positive outcome.

Conflict can sometimes be easy to spot. Symptoms of underlying conflict may include:

- Fewer people coming forward to volunteer for tasks
- Changes in the way people behave as individuals and towards each other
- People no longer working effectively and efficiently
- Sickness absence rates go up
- Open insubordination, even deliberate derogation of duties.

Who can be involved in conflict?

Conflict in the workplace may involve individual employees or a group of employees. Factions on ships may be based on ethnicity or race, religious or cultural beliefs, cultural backgrounds, gender or sexual orientation.

At the individual level, any of us may have a problem with one of our colleagues or a Master or senior officer. This can be caused by a clash of personalities, differences of opinion or values, personal issues being brought into the workplace, an overbearing management style, weak leadership or favouritism. Conflicts of these kinds are normal, but they need to be addressed to prevent them becoming serious, perhaps leading to a formal complaint of bullying or unfair treatment.

The dispute may be between individual crew members, different groups onboard or between employees and the management team. Factional conflict may stem from historical rivalry or the feeling that one group is not pulling its weight. Disputes involving the management are usually attributable to a grievance or to unhappiness with the management team's actions.

Conflict based on gender or sexual orientation needs exceptionally careful handling as it could put the ship in breach of flag state or port state laws. It can have a grievous effect on an individual, especially if that individual is in a minority.

Conflict arising from bullying also demands careful handling. The International Transport Workers Federation has produced an excellent handbook, www.ecsa.eu/projects/workplace-bullying-harassment, which is produced in several languages. It is recommended that all ships have a copy of this on board. There is a link to a useful video on the subject at www.nautinst.org/command

How do you deal with conflict?

The first step is to identify who is involved. Is the conflict just between two people who do not get on or is anyone else influencing either party? One of the trickiest situations to deal with is where an individual is being manipulated by another. It may be that two individuals seemingly have issues with each other but one party's standpoint is being reinforced by a third person. In these circumstances you must make everyone aware that such behaviour will not be accepted. You need to set the standard of behaviour, which is to treat everyone on board fairly and equitably.

Next you need to find out what is creating the conflict. Causes may include a personality clash, different needs/values, unclear job descriptions or division of responsibilities, poor communication, unresolved issues from past rivalry or a perception of favouritism.

Finally, think about why you need to address the situation. What benefit will it bring to you and the ship? What is the ideal outcome and what is an acceptable outcome? You may want to discuss the problem with a more experienced colleague.

Managing individual conflict

Conflict between individuals can often be resolved informally with a quiet word to the people concerned. Before you speak to them, consider carefully what is likely to be at the root of the issue and try to gather some background information. Consider asking another person to act as investigating officer.

Think about how they will benefit from changing their behaviours. You will need to consider how they may react and how you will respond to their reaction. A common reaction is denial. In that case, repeat the facts and give them time to reflect. Alternatively, they may express anger or distress, remain silent or seek to place the blame elsewhere.

Speak to them initially on an individual basis. Let them know you are aware there is an issue between them, ask them what that issue is and make clear the sort of behaviour you expect from each individual in future.

In the meeting, make sure:

- Enough time has been allowed for the meeting – it may take longer than you expect
- You have the background information you need

- You have examples of where the issue has been observed and potentially caused difficulties for the ship
- Your questions are open and direct
- People have an opportunity to talk while you listen
- You repeat what has been said to check their understanding
- You try to find common ground
- An accurate record is kept of the discussions
- That both parties thoroughly understand what is expected of them in future and that they agree to any actions or changes of behaviour needed.

When you bring the two people back together, give them a chance to speak, but quickly draw a line under what has happened and focus on the changes you expect from them. Ask them how they will make it work.

If you are not able to resolve matters, consider finding someone to act as a mediator to bring the two parties together. For this to work the mediator must gain the trust of both parties. Where it is not possible to resolve the issue informally you should seek advice from a senior manager and ask the HR team whether the matter should be dealt with under the company's grievance policy. The problem should not be left to fester.

Managing conflict between groups

The causes of factional conflict may well be rooted in history. Tensions may have developed over a number of years, for example when people from different companies are forced together after a takeover and have failed to merge into a unified team. Perhaps the two groups have different terms of employment, are used to working in different ways and have conflicting values. This may lead to a lack of cooperation and team working, and different working practices may make operations more complex. People may refuse to work with individuals from the other group and there may even be allegations of unfair treatment, bullying and harassment.

Once again, you will need background information if you are to understand the issue and find a solution. You may need to involve others, including those within the groups involved, to help find ways to get the groups working together. While this process is going on you will need to deal effectively and promptly with situations that arise out of these tensions.

Some key pointers:

- Decide which battles you want to fight and why
- Try to understand why the groups do not come together – put yourself in their shoes
- Think about the advantages for them if they make the changes needed
- Sell those advantages to them
- Make clear to everyone involved the standards of behaviour you require and ensure they all understand the consequences of not keeping to those standards.

You may have people who refuse to carry out work in the way that is required. Assuming that the reasonable and legal instructions are given by you or your officers, you are fully entitled to instruct them to carry out the task. Make sure that they understand what will happen if they do not comply. After the event you should explain why you require them to operate in a certain way and also discover why they objected to the request.

Avoiding conflict in the future

Nurture a positive workplace in which there is 'healthy conflict' by:

- Making sure everyone knows what is required of them in terms of their duties and appropriate behaviour
- Keeping people informed – make sure that they have access to information provided by the company and details of your planned voyages
- Listening to people – you can often pick up a lot over a cup of tea or from a throwaway remark (your daily 10 minutes on the bridge)
- Encouraging people to speak up and put forward ideas
- Leading by example, respecting each individual for their contribution to the ship and the voyage
- Exhibiting zero tolerance of unfair treatment
- Taking immediate and positive action to solve problems as soon as they arise
- Acting as role model, both as an individual and of the company values

Know when to get help from elsewhere.

By walking the walk and talking the talk you can create a positive, open atmosphere where people can bring forward problems to you, in confidence so that issues can be dealt with at an early, rather than a late, stage.

The basics of negotiation

By Captain Kuba Szymanski

As a Master there is a good chance that you will find yourself in a position where negotiation skills will come in handy. I wish I had possessed those skills when I was freshly promoted.

Examples of situations demanding skilful negotiation include:

- An unhappy seafarer threatening to leave the vessel because the company you both work for has not kept its promises on leave, salary or conditions
- The superintendent or office representative trying to impose unacceptable conditions
- A cargo interest representative suggesting inappropriate solutions
- An agent bullying you to take shortcuts
- Pilot or port authorities imposing unacceptable ideas
- A class or port state inspector, vetting inspector or auditor submitting a biased report
- Police, coastguard or customs making unjustified claims.

While far from complete, this list should give you food for thought. To deal effectively with problems of this kind you need to be prepared. Experienced Masters will consider 'what if' scenarios and rehearse good solutions. Take some time to reflect on these issues and your responses to them so that when you are faced with the situation in reality you are able to draw on those solutions.

Ask yourself three simple questions:

- Do I want to play it hard or soft?
- Do I know my best alternative to a negotiated agreement (BATNA)?
- What is my exit strategy?

Negotiating style

We all have our preferred methods when negotiating; some of us are very hard players and some are soft (sometimes too soft). Now ask yourself whether you generally:

- Try to avoid conflict
- Give in easily
- Are trustworthy and honest
- Focus on building relationships.

If these best describe your approach, you are a soft negotiator.

Alternatively, are you generally:

- Aiming for complete victory
- Willing to pressurise the other party
- Lacking consideration for the needs of the other party
- By nature a haggler.

If these are true, you are hard player.

As it is not easy to undertake self-assessment, you should ask someone you can trust and who will not be shy to give you an honest answer. Should you be lucky enough to have a mentor, this would be the ideal person to conduct this kind of appraisal.

Alternative scenarios

The second question concerns your best alternative to a negotiated agreement. This one sounds simple but can be hard to put into practice. As a Master, you may find yourself ambushed by these situations, and while it is great to have well thought-out contingency plans you may find that these are easier in theory than in practice. Unless you are already experienced you will need to have some premeditated scenarios in your head. You will have seen other Masters – your predecessors – in action, so think about how they approached these situations and perhaps draw some lessons from them.

To help you with this process, you could:

- Brainstorm a list of alternatives if negotiations lead to an unfavourable outcome
- Improve on more promising ideas and convert them into alternatives
- Identify the most beneficial alternative as a fallback

- Revise your BATNA as negotiations evolve
- Don't reveal your BATNA to the other party. Bear in mind that if your BATNA turns out to be worse than the other party had expected they may reduce their offer
- Start from the bottom (the least acceptable option for you) and go up the option ladder to the best achievable – 100% victory.

This exercise should not be rushed, so don't be afraid to ask for some time. When dealing with officials you can always request time to consult. The opposition's tactics may include pressurising you. Remember that they too have come with their BATNA, probably set pretty high, and they will try their options from top to bottom.

Finally, you need your exit strategy – it is extremely important that you stick to it. This is a point of no return in the passage plan for these negotiations. Once you cross it you are committed and you have no further choices. Make sure you don't forget your options and that you agree to the terms negotiated before you get pushed beyond this point. Otherwise, ask for more time and consult someone else straight away.

Useful hints

In some cases it is an advantage to have others with you; let someone else negotiate, leaving you to monitor the whole process and pull the strings. This might allow you to pressurise your counterparty. It might also be a better position from which to pull out, by being a saviour – stepping in when all hope appears lost. With two of you present one can play the bad guy, leaving you with the good guy role to play (or vice-versa).

Never, ever get angry; always stay calm and polite. When things become frustrating or your opponent is provocative, silently count to ten, stand up and offer coffee or tea; anything to help you break away from the discussion and give you time to calm down. Always respect your opposition and show that.

Wear uniform. You will be amazed what an official outfit can do for you.

As a final word I would like to reinforce the advice I gave you initially: think what if and get prepared. This is all about gaining experience.

Section 5

What if?

Accidents on board

By Captain Paul Drouin

There has been an accident on board and now you, as Master, must investigate. Why? To find a guilty party or to point the finger? Of course not. Accidents are investigated to learn what lessons can be learned from the experience. Investigations are a tool for continuous improvement, which is also one of the goals of ISM and quality management.

Almost all accidents, incidents and close calls are products of several contributing factors. Accident reports may cite carelessness or complacency as a cause of an accident, but this is not really correct. If employees are complacent, what about the supervision? Has the company or you, as Master, been turning a blind eye to complacent behaviour all along? If so, why has this gone undetected? In almost every case a thorough investigation will find one or more unsafe conditions that contributed to the event. Inadequate leadership, lack of training, tacit management acceptance of shortcuts – these are some of the common unsafe conditions often found. Of course, an employee who is careless or dangerously complacent should be dealt with accordingly – but this is rare. In your investigation, never stop at an unsafe act – there are usually one or more unsafe conditions lurking behind the event.

Before you can investigate, you have to be aware that something has happened. Of course, an accident or something major will be known to you as Master, but just as important are the many slips, spills and minor incidents or near-misses that happen frequently. In order to learn lessons from these incidents a robust reporting culture must be the norm. But reporting something that went wrong often goes against our natural instincts. Human nature is more inclined to hide such events than to broadcast them. No one wants to be the centre of attention, the butt of jokes or worse. Our survival instinct kicks in when we slip up or make mistakes and the preferred option is to say nothing, especially if there were no severe consequences. Since nothing bad happened, who will be the wiser and why tell anyone?

Research has shown that for every severe (personal) incident, sometimes referred to as a lost time accident (LTA), there are about 30 minor injuries and over 300 unsafe acts or actions. Although these are only statistics – numbers on a page – they speak volumes for the value of a strong reporting culture. These statistics suggest that there is at least one unsafe act happening almost every day. You must be aware of this and make great efforts to ensure that your senior officers are similarly aware: they supervise the daily working of the ship and must do so with safety in mind.

More and better reporting of minor injuries and unsafe acts will improve the overall safety situational awareness of the company or vessel. This heightened awareness should then provide a better vantage point from which to take decisions related to safety.

Although much can be gained from studying what went right, much more can be gleaned from analysing what went wrong. That is why a company and its senior leadership must instil a culture where reporting is not only encouraged but rewarded. In other words, reporting becomes a core value of the company and the crew. In such an environment all personnel view the reporting and investigation of an unsafe act or minor injury on a par with reporting a LTA or major casualty.

Establishing a reporting culture is not easy. It takes firm leadership and commitment to the principle of continued improvement. Some investigations are easier than others, some traumatic and others quite routine. But each accident, incident or close call is telling us something important – that levels of risk are not as low as it is practical to make them. Your job as Master is to find the unsafe conditions or underlying factors involved and apply sensible risk reduction strategies to improve safety.

The key is switching from a blame culture to what is increasingly being called a just culture. A just culture is one where human error is considered inevitable. In a just culture an organisation's policies and processes must be continually monitored, through a strong reporting culture and audit process, and improved to take those errors into account. Yet, a just culture does not condone complacency or negligence. Individuals should be accountable for their actions if they knowingly violate safety procedures or policies.

You should consider how best to instil this just culture and an essential element of this would be to encourage near-miss reporting. Many companies have a format for such reports and can demonstrate that their safety records have improved dramatically as a result. If your company does not have this in place, discuss with your technical managers why you would like to implement such a system and how this could be done. After all, you are now the Master on their ship and you have personal responsibility and accountability for what happens there – they should support your endeavours to minimise the risk of accidents.

Finally, help others learn from the incidents and accidents you have investigated. Send a report to the MARS editor at The Nautical Institute for publication in *Seaways* and inclusion in the MARS database that is freely available to all in the maritime industry. See www.nautinst.org/mars

Medevac – helicopter arrangements

By Captain Eric Patten

This article considers helicopter hoisting operations for medical evacuation (medevac) but does not cover landings on board. While not intended to address every situation nor be a substitute for sound judgement, it should help to familiarise the Master with the basic issues.

Helicopter operations to ships at sea are inherently dangerous and should be undertaken with great care. The responsibility for safe operation of the ship lies with the Master, while the safe operation of the aircraft lies with the pilot – each should be properly familiar with their duties. Wherever possible these operations should be conducted in daylight, as night operations are considerably more hazardous.

Hazards during helicopter operations

All personnel operating around the helicopter should be made aware of the unique hazards arising from helicopter operations.

- Rotor downwash – the 'wind' below a helicopter's rotors – can exceed 80 knots in speed and may lift debris, cause injury, damage equipment and blow personnel overboard
- Engine exhaust can be extremely hot and burn personnel and equipment. The risk is mitigated if the helicopter is hovering well above deck
- Rotor blades and tail rotors are a significant and deadly hazard to personnel
- Static electricity discharge from the hoisting cable/strop. Helicopters in flight build up static electricity. Before being handled, hoist cables, external cargo hooks and so on must be grounded with a grounding wand or placed on the deck
- Flash photography/laser devices should never be used during helicopter operations as they can disorientate the pilot.

WARNING: Do not handle the cable or cargo hook with bare hands before grounding.

Before operation planning

Planning for helicopter operations should be carried out regularly, using the *ICS guide to helicopter/ship operations* for guidance. As Master, you and your team should evaluate the hazards involved in all phases of shipboard helicopter operations and develop appropriate safety measures and procedures. You should ensure the correct equipment is on hand for the specific vessel.

Identify any equipment needed to safely conduct operations, such as:

- A grounding hook or wand to earth the hook and cable
- Personal protective equipment for crew operating underneath the helicopter. PPE should include head, eye and ear protection; gloves for handling the hook and cable; floatation equipment to protect a crew member blown overboard; survival suits, if necessary
- Day shapes, windsocks etc, as required
- An appropriate radio to communicate with the helicopter. Before starting operations, check whether the helicopter can communicate on marine bands. While operations are taking place the radio must be continually manned.

Ships should have an identified helicopter operations area that must be kept clear of debris and obstacles. All the weatherdeck areas, and particularly the designated hoisting area, should be inspected before, and monitored throughout, all helicopter operations to ensure that they are clear of moveable debris. Rags, pieces of paper, line/rope, hats, nuts and bolts and similar matter can be caught by downwash, causing damage to the aircraft or injuries to personnel.

WARNING: Prevent debris and unsecured items becoming airborne.

Shipboard personnel should be trained in safe procedures before starting helicopter operations. During operations, only essential personnel should enter the helicopter area; all others should keep clear or below decks.

Considerations

The Master should be aware of the following to ensure safe helicopter operations:

Before helicopter operations

- The helicopter rendezvous position should:
  - Be within range of the helicopter. Helicopters have limited range so it is vital to discuss the location with the helicopter operator to ensure that the operation can be conducted safely
  - Have adequate sea room for the ship to remain on a specific course and speed, with enough time to carry out the operation while remaining in range so the helicopter can return safely to base. If conditions permit, it is preferable to steer towards the approaching helicopter
  - Be clear of obstructions and hazards
  - Preferably have little or no traffic.

- Weather at the rendezvous point:
  - Relative wind is the most significant consideration for the helicopter, as the pilot will want to hover the aircraft into it. Relative wind is defined as the direction of movement of the atmosphere relative to an aircraft or an airfoil. It is opposite to the direction of movement of the aircraft or airfoil relative to the atmosphere. The pilot will want to approach the ship into the wind and then depart into the wind. Depending on the location of the hoisting area, this may mean that the ship has the wind coming athwartships or from the bow
  - The sea state must be calm enough to allow the ship to remain stable during these operations; it is secondary only to allowing the helicopter to hover into the wind. A compromise will often have to be reached between the wind and the ship's pitch and roll. Frequently, it can be achieved through the ship making more speed, and it may also change the relative wind
  - Cloud level and visibility may affect safe operations. Helicopters can fly very low but usually not in fog or in poor visibility. The ship should keep the helicopter operator informed about weather conditions
  - Where temperature and dew point are within 3°C the likelihood of fog is increased
  - Barometric pressure
  - Other weather – rain, snow, ice etc.

- Plot the rendezvous point and then a course and speed that will assist the operations. This will probably need to be hand-steered
- While helicopter operations are in progress the radio must be continually manned
- Display the correct special signals and shapes (ball-diamond-ball). If available, provide either a windsock or smoke and flags
- Brief all personnel about their responsibilities and the hazards of the operation. Make sure that they are properly equipped to operate safely.

During helicopter operations

- Communicate with the helicopter as early as possible. The pilot should be advised of the situation (passengers, patient status, etc.), location, course and speed, current weather conditions and the position of the hoisting area
- If necessary, help the helicopter find the ship. This can be done by radio homing, providing a description of the vessel (type, colour and configuration) and making white smoke
- Choose a safe course and speed for the operations. Be prepared to alter course and speed if the pilot requests it, and identify any hazards on that course
- The aircraft may either be in a high hover (30m+) to avoid obstacles and hazards on the deck or a low hover of 4-9m. The high hover is much more difficult and the operation will probably take longer than if the helicopter hovers lower. A highline rescue may be appropriate and the ship's crew should familiarise themselves with this procedure
- Once the helicopter has made its approach and is safely hovering above the hoisting area, the crew will lower the cable and lifting device (either a single strop or some sort of basket).

WARNING: To prevent electric shock, do not handle the cable or cargo hook with bare hands before grounding.

- The deck crew should ground the cable being lowered by the helicopter and acquire and maintain control of the hook/single strop/rescue basket using grounding wands. Procedures to ground a cable:

1. Connect grounding wand's ground clamp to a good metallic grounding path through ship's hull
2. Allow the cable to touch deck before contacting the cable with the grounding wand. Hook the cable with the grounding wand and maintain control of it
3. Once grounded, maintain continuous grounding contact until the hoist is retrieved.

WARNING: Do not attach the hook or basket to any part of the ship.

- In a single hoist, in which no aircraft crew member is lowered, the pilot's directions over the radio should be followed
- In a double hoist, a crew member is lowered to the deck. Care should be taken that the crewman is not injured when approaching the deck. Once on deck the ship's crew should follow directions about hoisting to the helicopter.

After helicopter operations

- Maintain communications with the helicopter as required. They may need you to relay information to shore via other means such as satellite phone
- All equipment and gear used in the operation should be inspected for damage and wear. Repair and replace as necessary
- Stow all gear
- Debrief crew.

Now we consider the arrangements that would need to be made to help the operation by those ashore.

Medevac – onboard arrangements

By Captain Pushkar Gadam

Medical evacuation at sea (medevac) can be carried out by transferring the patient either to a helicopter or aircraft or to a vessel's lifeboat. The mode of transfer will largely depend on the vessel's distance to the nearest port or airport. Depending on the type of injury or illness, speed of evacuation may be important for saving the patient's life. During such operations it is important to keep calm, maintain vessel discipline and avoid endangering the rest of the crew or the vessel.

Evacuation by helicopter is becoming increasingly commonplace, but is limited by the type of helicopter available and its range. The main advantage is that helicopter crew often have a medic in their team who can be dropped or landed on the vessel to carry out an initial diagnosis, give advanced first aid and stabilise the patient before evacuation. Masters should note that the patient may still feel anxious and uncomfortable in a helicopter because of the movement, vibration and loud noise. Refer to the previous article on helicopter rescue if you find yourself in the position of needing this type of medevac.

Vessel-to-vessel evacuation is more usual when the vessel is close to the coast. Vessels used are small and fast, but their range from the coast is limited and affected by weather conditions. The freeboard of your own vessel and personnel transfer arrangements will also have to be considered. For both of these options, a proper risk assessment should be carried out before the medevac operation.

Following a medical incident onboard, the first point for guidance will be the World Health Organization's *International medical guide for ships*. The Admiralty List of Radio Signals (ALRS) Volume 1 provides contact information for the International Radio Medical Centre (CIRM) and there should be no hesitation in seeking advice if the circumstances require. Masters can also contact the nearest coast earth station (CES) or coastguard station, details of which are also provided in ALRS Volume 1.

The decision to carry out a medevac and selection of the mode of transport should be concluded after close consultation with doctors and the organisation that will conduct the evacuation. The Master should prepare to take part in a three- or four-way conference call that includes the doctor, coastguard and the vessel. The coastguard will set this up.

The Master should ensure that information is sent to the correct external recipients, as some companies have a confidentiality protocol for such incidents. The patient's next of kin must be informed of the situation, and the technical manager should be explicitly requested to do this. Details of such information may be found in the crew contract.

All events of this kind must be properly logged. The vessel owner, in conjunction with the vessel's P&I club, flag or coastal state or local police, may well undertake further investigations into the incident.

In complex cases, the vessel's P&I club can provide assistance to the Master. Its local correspondents will have very good local knowledge and contacts with local authorities. Masters should note that the P&I club representatives are working for their benefit and will provide as much information as possible. Some P&I clubs, however, will require early notification of the incident in order to carry out an independent claim investigation. The vessel owner should take care of this.

While all this is being organised, it is important that the patient is always attended to and kept informed, if conscious. If unconscious, the patient should be placed in the position advised by the Maritime Telemedical Assistance Service (TMAS) and vital signs continually monitored. It is important to collect together all relevant information and to ensure this accompanies the patient. This may include, but is not limited to:

- Passport, seaman's book, vaccination book
- Medical reports, which should include as a minimum:
  - details of onset of illness or injury
  - description of medical findings and a log of their development over time, especially vital signs such as blood pressure, breathing frequency, Glasgow Coma Scale etc.
- Therapeutic measures taken, especially medication administered
- A printout of all correspondence with a TMAS or with doctors in previous ports concerning the patient.

The patient's personal belongings should be packed and sent along with them, if this is practical and time permits.

Depending on the incident, crew or passengers may feel vulnerable, so appropriate measures should be taken. After the incident and following successful medevac of the patient it is advisable to call a crew meeting to discuss the matter openly. Details of the hospital to which the patient has been sent should be requested from the coastguard station for this debriefing. It is important that this debrief takes place, because fellow crew members may be traumatised by the incident and need further help and reassurance.

Dealing with an oil spill

By Captain Alex van Wijngaarden

The response to shipboard oil spills is far more prescriptive than for other shipboard emergency plans which were developed and refined through onboard exercises, drills and experience.

Oil spill response is set out in Annex 1, Regulation 37 of the MARPOL Convention. The requirement to prepare approved shipboard oil pollution emergency plans (SOPEP) applies to all tankers of 150 gross tonnage or greater and to all other ships of 400 gross tonnage and above. The plan provides valuable guidance for the prevention of, or dealing with, shipboard oil spills. It also outlines the requirement for the Master to report any actual or probable spills to the nearest coastal state as quickly as possible.

It is generally beyond the ship's resources to respond to oil spills that occur as a result of a grounding or collision. Should an accident like this occur, the authorities will seek information, as a matter of priority, on the tank or tanks that have been assessed as damaged and the nature of the product contained in them. One of the first actions to be considered by the Master is to reduce or minimise the outflow of oil by transferring fuel from the damaged tanks. Authorities are likely to insist on this. Depending on the severity of the damage, the classification society should be advised as soon as possible so that any damage stability issues are considered.

The majority of spills generally occur during usual shipboard activities, often during bunkering or internal fuel transfer operations. Operational spills can have a wide variety of causal or contributory factors. This could include mechanical (or other) equipment failure or mistakes happening as a result of inattention, fatigue, inexperience or a combination of these. Many jurisdictions will have requirements that must be observed before taking or transfer of bunkers within the ship. Many of these will require permission from the Harbour Master, or someone in a similar position of authority. The ship's agent is best placed to provide guidance on this.

Irrespective of the cause of the spill, there is some basic information authorities will seek almost immediately: product name; specific gravity; viscosity; pour point; solubility; flash point and, where available, the material data safety sheet.

Accurate information on the amount of oil spilt is also vital. It may be tempting to under report but the true amount, if known, should be provided to shore-based personnel as soon as possible. It quickly becomes apparent if there is an effort to downplay the amount spilt. This type of behaviour results in the credibility of the Master being called into question, creating an atmosphere of mistrust of any subsequent information provided by the ship. If the amount spilt is unknown, provide all available information, giving a clear indication that the volume is unconfirmed. Make every endeavour to provide an accurate amount as soon as possible – even if it is a best estimate. This information will assist shore-based response teams in determining the best method of responding to the spill.

Any spill is certain to result in quick and decisive action from the authorities. It is essential that those on the ship take any action they can to reduce or stem the flow of the spill to mitigate the potential impact. Such action may influence the subsequent approach of the authorities. For example, a quick response to immediately activate the SOPEP plan is far more likely to lead to a cooperative response from local authorities. It is also the best form of mitigation for the ship should the incident result in court action.

A shore-based response mounted by authorities will be based on a tiered system ranging from tier 1 through to tier 3. Tier 1 response is implemented for smaller spills. At the other extreme, a tier 3 spill is one where the magnitude potentially overwhelms local responders which then require the support of nationally coordinated resources. In addition to setting out the operational aspects of a response, the local contingency plan also identifies environmentally sensitive sites that may potentially be impacted by a spill. Until the authorities have an accurate assessment of the spill, Masters should not be surprised if the initial response appears to be excessive. This will very quickly be adjusted to meet the response requirements.

Response options generally consist of on-water boom containment and recovery, application of approved dispersants, shore-line clean up or a combination of these. On rare occasions when it is clear that the spill is not likely to threaten resources and is moving offshore, monitoring only may be required.

The response does not necessarily terminate once all visible oil has been removed. It is not unusual for monitoring programmes to be initiated to ascertain the impact of the spill on the environment but this will depend on the size of the spill and the type of oil spilt.

If there is a spill, contact with the technical manager and the local P&I correspondent is essential. The latter in particular will have existing relationships with authorities and be able provide the necessary advice on dealing with them. Consequences of a spill vary from jurisdiction to jurisdiction but the spilling of oil will invariably constitute an offence. Limitation for the laying of charges will vary, but where charges are laid, the penalties imposed may be financial and may also include incarceration. Some jurisdictions may insist that the Master remains at the port until completion of any judicial process. In addition, most jurisdictions have adopted the 'spiller pays' principle, firmly placing the cost at the technical manager's door.

As with everything, prevention is the best cure. The risks of a spill can be mitigated by ensuring that there are robust and well defined procedures in place (that are followed). Bunkering and internal fuel transfers should be conducted under the supervision of a suitably qualified person. Procedures should include authorisation of internal transfer/bunkering, supported by comprehensive checklists that are well tried, tested, accepted and used by the relevant staff.

As with any emergency response plan, the SOPEP must be regularly exercised to ensure that crew are prepared to respond effectively and in accordance with the plan. All crew members need to be thoroughly familiar with the emergency procedures, principles and objectives.

Masters must be aware of the potential consequences of the loss overboard of any dangerous goods carried, either in package form or in bulk, as many are classed as marine pollutants. This will also count as a pollution incident. You and your crew need to be familiar with the appropriate sections of the IMDG Code that relate to these products, as well as with the MFAG section, and be prepared for any emergency.

As Master you must be aware of the consequences of any pollution incident. It could lead to detention of your ship or detention and even imprisonment of you as Master. Therefore ensure that all officers and crew are aware of the implications of a pollution incident and mitigating factors and are well trained. Impress on them the very serious results of trying to hide anything from the authorities.

Emergency response

By Captain Alex van Wijngaarden

An emergency can occur at any time, while at sea or in port and the response to it will dictate success or failure. Irrespective of the fact that you are in command for the first time, the person the crew will look to for answers is you, the Master. This is where the buck stops.

In an emergency you will be dealing with some very complex situations onboard your ship and you need to be aware that the effectiveness of the overall response will be governed, in part, by the nationality and culture of your crew. Multi-national crews are now the norm and, in the event of an emergency, individuals under great stress may revert to their mother tongue and not respond effectively to the language commonly spoken on board. Similarly, culture will have an influence on how individuals react when faced with emergencies. Cultural influences may be such that individuals may not initiate actions, and will simply wait until directed.

As Master, you need to get to know your crew and try to understand and appreciate their strengths and weaknesses; in an emergency, that knowledge will allow you to better manage your people, as well as the emergency itself.

The Master's primary role in an emergency is to maintain the 'big picture' approach. The chief officer's role is direct involvement in dealing with the emergency, whereas the Master's role is to stand back and coordinate the activities necessary. If you get involved in the operational side of any response you risk losing situational awareness, and potentially losing control of the situation. It is by no means easy to display a calm presence in the face of an emergency but this is what is required. Those around you will very quickly pick up on doubts and anxieties displayed, even sub-consciously, through body language.

The key to the success or failure of response to an emergency will depend largely on the preparatory work undertaken in planning for the many emergencies that may occur on board.

Contingency plans cannot possibly cover all scenarios, but can provide guidance for a range of possible scenarios. The Master should bear in mind that these contingency plans have been prepared to provide bullet points but it is important to remember that they cannot and do not provide all of the answers.

Many of these plans are already likely to be in place, such as:

- Damage control
- Fire plan
- Shipboard pollution emergency plan (SOPEP). See pages 114-116.
- Recovery of people from the water
- Towing plan
- On passenger ships, a decision support system plan for the Master.

A good plan does not need to consist of lots of paper – it should cover all the necessary aspects of a particular type of emergency and the broad approach to be taken. Take the time to go through these contingency plans and test their appropriateness to ensure a clear understanding of their content.

The Master needs to be satisfied that the relevant response plans will be effective; to do so, you should ensure that they are exercised on a regular basis so that the crew becomes thoroughly familiar with them. This will take patience and perseverance and provides a chance to review the currency of these plans. Additionally, providing variety in exercise drills maintains crew interest and commitment. See pages 63-66.

Encouraging full crew participation, identifying the learning outcomes expected and making sure crew recognise these, will ensure crew members have a clear understanding of the response plans when a real emergency does occur. This will clarify the tasks expected of them and also serve to highlight the need to operate as cohesive teams. Key to this is motivating the crew to approach each exercise or drill as if it were a real event, through making drills and exercises interesting and challenging.

Remember the Five Ps – proper planning prevents poor performance!

No amount of planning and exercising will prepare the Master for the time an emergency really occurs. When the alarm bells start ringing, being able to make decisions based on logical reasoning that lead to a solution to the emergency is essential for a positive outcome of any response.

This forms the basis of the Appreciation process, developed to solve military problems. In summary:

Aim	A statement of what has to be done
Factors	Relevant facts influencing how the aim will be achieved
Courses	Options available to achieve the aim and weighing up advantages and disadvantages of each option
Plan	Best course of action which develops into an action plan.

Although this may seem a long-winded process, in reality it will take a few minutes of thought, allows for the ordering of these thoughts and provides the leadership and direction the crew expects.

Emergency response – spiral to success

Instructions:
1. Follow each stage logically
2. At stage 4, if the answer to the "Am I winning?" question is 'No', return to stage 1
3. Follow each stage logically until the answer is 'Yes'

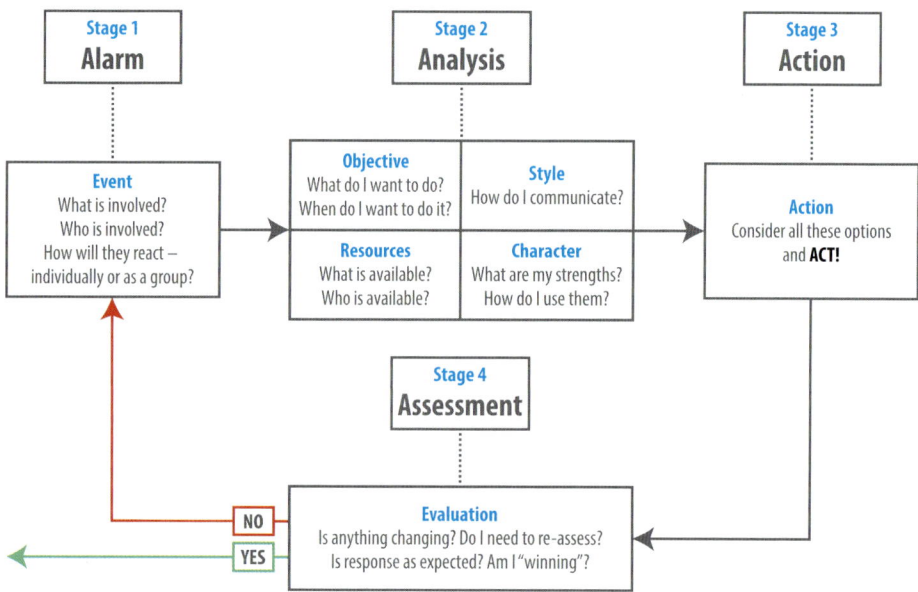

The flow diagram has been developed and adapted from many sources and will provide a sense of direction and purpose to the management of any emergency. You cannot move from one stage to the next without ensuring that you have all the relevant information. However, you must not assume, because of a lack of feedback, that your action plans are effective. If Plan A is not achieving what you expect, try to find out why. Reassess the situation and move to Plan B, or C or D, if needs be.

Whenever an emergency occurs, whether at sea or in port, the crew's familiarity with first response plans is key to a successful outcome. If the vessel is at sea, assistance may still be days away – or it might be close to hand. Establishing early communication links with the nearest coastguard or Coast Radio Station is vital as they will be able to co-ordinate assistance. Also, there is no harm in broadcasting a Pan-pan or Securite message on VHF Channel 16: there may be other ships in the vicinity able to offer assistance.

However, until such assistance arrives (if it is indeed available), it is only through exercising vital components of individual plans that the crew can realistically be expected to mount an effective response. For ships carrying dangerous goods, either in package form or in bulk, the Master must be aware of the potential consequences of the effects on any of these cargoes by fire, flooding, impact and the like. Many of these goods are classed as marine pollutants and you and your crew need to be familiar with the appropriate sections of the IMDG Code that relate to these products, as well as with the MFAG section, particularly if your emergency teams are exposed in one way or another to these cargoes.

In port, on the other hand, there are shore-side resources capable of providing support to the Master and crew in dealing with an emergency. The key here is 'support', the Master always remains in command and responsible. Shore emergency response organisations generally only have limited knowledge of ships – for example the potential consequence of using large amounts of water to extinguish a hold fire. Additionally, these organisations may not have a mandate to respond to shipboard emergencies. Local Harbour Masters will be able to provide the main link between the Master and shore response organisations in most instances and, with their maritime background, can ensure that the response organisations provide the correct level of support.

Emergencies can occur at any stage of the voyage but having a crew that is capable of responding to them with skill and confidence will result in a successful outcome.

Criminalisation

By Captain James Robinson

Criminalisation is a problem for mariners and in most cases it is the Master who is seen as the criminal. All seafarers need to be aware that in the event of an incident, particularly one involving pollution, which can stir up a very emotional response in local communities, you could be subjected to criminal charges. However innocent you think you may be and however obvious the cause of an incident may seem to you, local legislation will apply and local media and politicians will almost certainly view the incident in a different light from that of the ship's Master and crew.

If seafarers do everything in their power to avoid an accident or to mitigate the consequences of an accident, and they can demonstrate that they have done so, then they should not be penalised. However, in many cases we see ships' crews being held as scapegoats and used as bargaining levers in the hope of obtaining more compensation than might be legally expected. A coordinated approach by representative bodies of the maritime industry is essential to combat this.

What can you, as the Master of the ship and as a professional mariner, do to protect yourself and your crew from criminalisation?

Section 5
The Nautical Institute on Command

While you cannot control all the possible contributing factors that go to cause an accident, you should try to control as many of them as you can. The first thing that you can control is yourself, by educating yourself to do the best job that you can at every level of your career from cadet to Master and beyond. By being a competent mariner you are reducing the risk of being involved in an accident. Take every opportunity to enhance your own competence and professionalism through constant study, experience and continuing professional development (CPD).

Ensure that all the officers and crew under your command are very competent and aware of their responsibility towards the safe running of the ship and the possible serious results of an incident or accident.

Follow the guidelines and recommendations given by others in this book with regard to training, to emergency exercises and response. Monitor and mentor the activities of your officers and crew and encourage them to challenge your own decisions and actions.

By working together, you will run a safer ship and, in the unlikely event of a major incident, you will all be well prepared in your responses and able to show that you have all done as much as could have been expected of you. Unfortunately we live in an age of easy litigation and all seafarers must be aware of the consequences of any accident or incident, however small.

The Nautical Institute continues to challenge the increasing practice of criminalising seafarers and it reviews the Master's role and responsibilities to take account of changes in industry legislation and practices. At the IMO, our work continues to encourage the ratification of conventions and their implementation in national law to protect seafarers from criminal proceedings.

Members of The Nautical Institute frequently report to us their concerns about the perceived unjust criminalisation of seafarers involved in accidents and incidents. Unfortunately, many coastal states have brought in legislation that imposes criminal sanctions for many different types of maritime incidents and accidents and this appears to enable foreign governments and administrations to criminalise foreign seafarers with impunity. The industry is an invisible one in many countries – few people have a direct connection to the industry and fewer still give shipping a second thought.

The Nautical Institute believes it is essential that maritime administrations observe and adhere to their international responsibilities under the United Nations Convention on the Law of the Sea (UNCLOS) including article 230, which bars states from imprisoning seafarers serving on foreign-flagged vessels except in cases of wilful and serious acts of pollution within their territorial waters. Unfortunately while most states are signatories to UNCLOS, not all are.

At The Nautical Institute we support those seafarers unfortunate enough to be involved in criminal cases in the course of their professional duties. One of the valuable membership benefits is legal defence insurance cover to assist members should they require legal defence. We also offer practical support to members caught up in cases involving injustices and unfair treatment by assisting their defence lawyers on aspects of the case within the Institute's professional competence. If you become involved in an incident and you are a member of The Nautical Institute, let us know as soon as possible so that we can assist you in your defence.

If you are reading this and you are not a member, we strongly encourage you to consider joining us.

At The Nautical Institute we continue to work with the industry to review data on criminalisation cases, identifying the unjust and publicising the injustice of such cases, including at the IMO. We must all promote and encourage the worldwide application of the IMO's guidelines on fair treatment of seafarers in the event of a maritime accident and work with others in the industry to raise the profile and recognition of the maritime professional and of the maritime industry in general.

By ensuring that our industry is more visible to the public in general and by publicising and highlighting any unjustified criminalisation that becomes apparent, pressure will be put on maritime administrations to be sure that any prosecutions they undertake are fair and reasonable.

Crime at sea

By Steven Jones

Crime at sea is one of the threats to seafarers, vessels and cargoes that has to be addressed by onboard security arrangements. However, the risk cannot be completely eliminated and it is important for Masters to know what to do if the worst should happen.

If we take crime to mean unlawful acts punishable by states the list of crimes that may be committed onboard would be long, including, but not limited to:

- Arson
- Assault
- Bribery
- Cyber attack
- Embezzlement
- Espionage
- Fraud
- Murder
- Mutiny
- Piracy
- Rape
- Sabotage
- Smuggling
- Substance abuse
- Theft

- Vandalism
- Violence.

It is possible that many Masters will not have to deal with any of these crimes during their careers but you need to be aware that they may happen. In the unlikely event that they do, there should be plans in place to minimise the risks, mitigate the impact and detail the response to be implemented. A systematic approach is needed.

Criminal acts can be committed on all ship types and it is important to consider what should happen next. Such acts will need to be managed in a considered, legal and effective manner, while there are also issues of evidence collection, investigation and jurisdiction.

Guidelines

Where crimes occur it is vital that evidence is collected, records made and evidence safely and properly saved. The IMO has developed guidelines to assist Masters, based upon IMO Resolution A.1058(27) and prompted by serious crimes such as sexual assaults taking place on ships and cases of missing persons, particularly from cruise vessels.

The primary purpose of the guidance is to assist in preserving and collecting evidence following an allegation of a serious crime having taken place onboard, or following a report of a missing person, and on the pastoral and medical care of victims.

As Masters have responsibility for their vessels, the duty to conduct inquiries in order to collect evidence in connection with any offence falls to them. You should collect and document as much evidence as possible and collate a report in an agreed format. See *The Mariner's Role in Collecting Evidence*, published by The Nautical Institute, for further details.

Whatever the ship type, the key period for the Master is that between the report or discovery of a possible serious crime and the time when law enforcement authorities or other professional crime scene investigators take action. It is recognised by the IMO that Masters may be called upon to collect evidence that may otherwise be lost if no action is taken, but this is an exceptional situation.

The IMO guidelines focus on what it is possible to do onboard to preserve and collect evidence. Masters must be able to respond appropriately, despite not being expected to act as professional crime scene investigators, and ensure that all relevant information and evidence is gathered for any subsequent trial or investigation. Should any evidence be missing or damaged during the investigation, the guidelines do not establish a basis of any liability, criminal or otherwise, of the Master and officers in preserving or handling evidence.

The guidelines also contain directions derived from MSC.1/Circ.1404 Guidelines to assist in the investigation of the crimes of piracy and armed robbery against ships, including procedures on recovery and packaging of evidence. There are also suggested formats for victim statement, alleged perpetrator statement and independent witness statement.

Collecting evidence

The Master and officers working on the crime scene must understand the importance of processing it properly. The key elements in crime scene processing are:

- Appropriate collection and preservation of the evidence
- A clear record of everything that has happened to the evidence from the moment it is found until handed to the appropriate authorities.

Evidence should be collected in a way that prevents loss or damage. It is a good idea to have a plan for collecting evidence, as well as the tools to do so. Masters are advised to develop a crime scene toolkit – a box that contains all the equipment, checklists and means of recording evidence they will have to use in the event of a crime onboard.

Fair treatment of seafarers

Fair treatment of seafarers is an important consideration where potential crime is concerned, bearing in mind the recent trend of criminalising them. Proper and effective evidence gathering and recording can be important in protecting crews. See pages 119-121 for more on criminalisation.

Who will have jurisdiction?

A thorough investigation of a serious crime onboard may be a lengthy process and it may not always be certain which state or agency will take the lead. Where more than one state may have jurisdiction in the case, this is likely to present further complications and challenges to the authorities responsible for the investigation.

Discussions on jurisdiction are likely to be beyond the control of the Master or company and the provisions laid down within the United Nations Convention on the Law of the Sea (UNCLOS) will come into play.

If a crime does occur onboard your ship, there will be many challenges for you to manage – and you will need to deal with them in a calm, collected and methodical way. As well as gathering information and collecting records you must ensure that the company, insurers, lawyers and relevant authorities are fully briefed.

Dealing with death on board – the emotional impact

By Lynda Bailey

What do you do when someone dies on your ship? Do you deal with it as just a practical problem, or do you also consider the emotional and ongoing effects of everyone on board? I would just like to put forward some points that you need to consider, particularly with a multicultural mix of crew.

People respond differently to violent and accidental death than to death by illness or natural causes. Suicide and death involving blood and mutilation will evoke strong reactions of fear, disgust, horror and disbelief. It may bring up memories of experiences that reinforce past reactions, invoking the fight, flight and freeze responses.

For some people who have experienced many deaths, this one may be the final straw, the past experiences piling up to cause overload and an inability to deal effectively with their feelings, thoughts or behaviour.

If a person was present at the death, whether peaceful or violent, it may cause flashbacks, anxiety, panic attacks or depression. If your crew are behaving oddly, eg zoning out, being aggressive, avoiding contact with others, check that they are not suffering from post traumatic stress disorder (PTSD).

Bearing in mind that grief is not a linear process, most people will roughly go through:

Shock/denial – they will feel numb, disbelief and disorientated, but functional. It's important to remember that your crew may appear fine for a number of weeks or months before they move into the next stage of:

Strong reactions – they may feel anger, rage, envy, resentment, guilt and bitterness, followed by fear of pain or death, being alone or rejection. They may begin to act out these feelings with others, becoming aggressive, clingy, sullen or argumentative. This may be followed by:

Withdrawal – depression and displays of extreme grief; crying, sleeping a lot, not showering, talking obsessively about the event. This is when you will notice they are not coping with their work or relationships.

This process can be helped enormously by addressing each stage appropriately.

By providing a safe environment, where all participants can talk about and make sense of the event, you can go a long way to avoiding the development of PTSD. Some may need one-to-one counselling and the seafarers' welfare organisations, such as the Mission to Seafarers, may be able to provide a chaplain or counsellor to visit the ship. We have provided links on The Nautical Institute's publications website for these and other support organisations. See www.nautinst.org/command

Narrative coherence is the process of making sense, moment by moment, of an event. This allows everyone involved to put together a working model of what happened and addresses their fears, tangible and intangible, so that resolution and closure can be achieved. This allows the brain to store the event as a past experience and avoids the development of PTSD.

Remember that a person can move in and out of these stages, even return to an earlier stage, before reaching the final stage of acceptance and resolution. Grieving has no time limit and the person who drops into extreme grief a year after the bereavement will be just as surprised as you are. It's really not helpful to suggest that 'they should be over it by now'.

Many religions have complex and strict rules on who may touch and wash the body, the timescales for washing and burial or cremation, what should be done with the body, what position it should be placed in, returning the body to where it was born and whether post mortems are allowed, to name a few. You need to demonstrate your understanding of these cultural and religious requirements. These rules affect the immortal soul and, therefore, affect all members of the deceased's religion on board.

Imagine for a moment that your younger brother has died and you believe his immortal soul will be lost if he is not returned to his place of birth within three days? Although you understand intellectually this is impossible, the thought of him lying in the freezer for a week until you reach land is emotional torture. What will your parents, community, and religious leaders think? Will they hold you responsible? The guilt may be overwhelming.

As Master, what can you do? Forewarned is forearmed. Although you may not be able to change the practical and legal procedures for dealing with the body, you can have information at hand that allows you to address expectations as sensitively as possible. Ask the question "What, within the bounds of practicality, can I do to support your needs?" Can you accommodate them by allowing family members to wash and position the body, if they are available? Is there a member of their religion onboard who could perform some of the rites? It may be that there is nothing you can do, but by showing respect and understanding you will at least show that you care, and maybe, avoid some of the misunderstandings that can arise.

So, it all comes down to assumptions.

Don't assume people will cope because they aren't close to the deceased.

Don't assume that family and friends will suffer more severely, although that might be the case. It very much depends on the quality of the relationship with the deceased.

Don't assume that people will be 'over it' within weeks/months/years. Some people grieve for decades.

Don't assume you know what people need or want to help deal with the loss – ask!

Dealing with death on board – the practicalities
By Captain Alexander Sagaydak

What can you do at a time like this? It will be a time of great stress for you but you will have little time to consider your own feelings. Your crew will be looking to you to provide leadership so your first action will be to bring yourself under control.

In all circumstances carefully search the area where the body is found – any detail may help with the possible investigation. In the case of death due to illness be careful; avoid direct contact with the body by you or the crew. Pack the body into an airtight plastic bag in case of infection. Follow any advice from the shore doctor on how to proceed.

Make an inventory of all the person's personal belongings in the presence of at least two witnesses, ideally taking pictures. Your description should be detailed; for instance, with mobile phones, make a note of the type, and any particular details, such as a scratch on the back. This could be very important – relatives will be very sensitive, and even small items could mean a lot to them.

All documents, including licences, certificates, passports and seaman's book must be listed separately giving the full name of the document and the number of pages. Be especially careful with the Seaman's Employment Agreement; in some cases this could be the only document supporting any compensation claim. You need to also look out for the health list and/or other medical certificates.

You must be vigilant about the psychological health of the rest of the crew. They may want to discuss the incident endlessly and there may not be any other topic of discussion in the messroom. Your duty is to prevent the spread of rumours on board, including any discussion about level of compensation or the cause of death etc. Such matters can have a deleterious effect on the atmosphere on board and could be harmful to the reputation of the dead person. See pages 123-125 for more advice.

The body will have to be stored and there is no better place for that than the freezer. Clear it of food and put the body there. This is one issue that many members of the crew will struggle with. This is the thought that they are on the vessel with the body of someone they met every day and worked with. Be aware this is potentially a great cause for stress for everybody. You have to find the right words for your crew in this case.

Explain the situation to the crew, prefacing your talk with a few good words about their late colleague. Be clear about any actions you require from the crew. If it is possible, have a pause in routine so they can consider issues of life and death; but the best way to combat depressed thoughts is to keep the crew busy. Change the ship's routine accordingly: there is always plenty to do on board.

You need to consider how you will respect the traditions and religion of the dead person. You may not know specific traditions, for instance whether to use dark or white for mourning, but you can consult with the friends of the dead person or with the manning agent. It is impossible to follow all the traditions on board, but you may be able to apply some.

Pay attention to the dead person's close friends as they might be in shock. Don't hesitate to contact a shore-based psychologist if you see any signs of problems – you don't need any more trouble on board.

The death must be reported to the shipping company and other interested parties including the flag state, agent and the manning office. Then you must wait until you reach port.

Delivery of the body to the port state authorities prompts a whole range of procedures. These may vary from state to state, but some basic items are similar: a specially appointed doctor, sometimes called a coroner, has to prepare medical documents about the cause of death.

Of course the body cannot be just discharged ashore; your duty is to give your last respects. The local agent should help in these arrangements. Consider a special ceremony on board or on the pier tailored to the cultural and religious traditions of the dead person.

Most regimes will demand a post mortem as the death would be considered unexpected. Usually the authorities will involve the police, which may include collecting of evidence and questioning the crew. The ship's agent will collect all documents from the authorities and deliver them to the ship or to the ship manager.

Until the death certificate is issued repatriation arrangements for the body cannot start. The dead person is unlikely to be from the port country. The country of repatriation may only accept death certificates it has issued. An application will have to be made to the embassy of the country of citizenship by the agent, ship manager or manning agent. This is usually a long process.

It's possible that country doesn't have representation in the country the vessel is berthed in. In this case all documents, such as the local death certificate and investigation report, will have to be couriered to the nearest embassy.

Documents needed to complete quarantine and customs formalities include:

- Death certificate
- Document stating the cause of death
- Certificate of embalming
- Certificate of absence of infection
- Certificate of absence of enclosures
- Quarantine certificate
- Certificate from sanitary authorities, stating that transportation is allowed.

In some cases all these will need to be officially verified or notarised by the embassy.

One more piece of advice: don't try to be a messenger, unless you know the family very well. Leave it for the company's representative.

I hope you will never use these notes, however… memento mori – Masters should be ready for everything and if such disaster does happen – I hope this advice will help you.

Dealing with the media – professional and social

By Steven Jones

Your company should have a policy for dealing with all media, including social media. As Master, your only contact with the professional media is likely to be if your vessel is involved in an incident, so it would be a good idea to know beforehand what is in that policy and who is responsible ashore.

In the event of an incident, the policy will aim to ensure that:

- The company (or manager) is seen as the authoritative source of information
- Information is accurate
- Information is updated as facts emerge.

Your immediate responsibility is to give your company (or manager) the full facts (and only the facts). Do not try to hide anything substantive, even if it is damaging. Do not speculate. Instruct officers and crew that they are not to talk to the media.

In the event of any incidents or accidents, the social media/online output from a vessel can be hugely significant. Images can rapidly emerge on the internet long before any authorised release. Lawyers, investigators and the relevant authorities will interrogate social media sources in the event of an accident or incident, so postings can be hugely damaging to both seafarer and ship operator alike.

Even when there is no incident, posts on social media have the potential to reach thousands of people and to damage the company's reputation. Your crew need to know what they should and should not do and you, as Master, should not be burdened with monitoring what they are doing.

Example of social media policy elements:

Do…

- Know the company's guidance and principles
- Set your profile settings so only those you want to see, can see
- Stop and think before posting
- Imagine your colleagues or superiors can read the post
- Remember what you say cannot be unsaid
- Use common sense and courtesy
- Be respectful of other cultures, religions and values
- Respect copyright
- Monitor responses to your post, and ensure they are true, legal and respectful
- Report to the Master if you find any harmful or unpleasant comments.

Do not…

- Be foolish, naïve or act without thinking
- Be insulting, intimidating, threatening or embarrassing to others
- Post defamatory, obscene or threatening materials
- Share internal private company information
- Comment on your company's business practices
- Cite colleagues or post without their approval
- Risk your job or those of your colleagues
- Ruin your reputation or that of your company or colleagues.

Connectivity is a key part of having a motivated, engaged and satisfied crew, so ignoring the issue of social media or prohibiting internet access is not a long-term, sustainable answer especially as the ship is the seafarers' home. Companies need to ensure that it is managed to best advantage for all.

However, increased connectivity can open a vessel's onboard systems to potential cyber threats, such as seafarers unwittingly downloading a virus or some similar danger to computers onboard. Cyber security is a serious issue which must be managed and the risks mitigated against.

Seafarers need to be made aware of:

- Potential implications of their actions
- Sanctions for any wrong doing
- Actions which may prompt a negative response.

Crew should be aware that information posted online is public and may be viewed by colleagues, clients, competitors, the authorities or the media. All employees have general obligations to act in the best interests of the company, and not breach company confidentiality or the relationship of trust and confidence that exists.

A code of conduct is an excellent starting point in the development of best practise. Very few individuals would set out to knowingly cause problems, so ensuring people are aware of the potential issues can be hugely beneficial and will reduce the likelihood of problems. As with policies on safety, the environment and security, make sure these have been read and understood by all.

In developing a code of conduct, consider:

- Any vulnerabilities and steps to mitigate risks
- Steps to define acceptable behaviour
- Integrating social media with other management policies
- Crisis planning and management.

Such a code should provide a guide to the values, behaviours and ways of working which are expected of an employee. There can be legal considerations when posting online, and flag states will probably have to address issues of jurisdiction eventually.

Guidance should encourage personnel to ask for advice if in doubt, and they should be constantly reminded that there are always consequences to what is published online.

Sometimes seafarers take to social media if they feel concerned about issues such as safety or conditions onboard. While it is legitimate to raise concerns, personnel should know the correct and preferred route to voice their disquiet in the first instance. For this reason it is vital that companies provide a proper and effective reporting system. It can also be beneficial for there to be an intranet or internal "chat" alternative, so seafarers can share their thoughts internally and still feel connected but without the damage which can come from external scrutiny.

Masters should not have to constantly monitor outgoing communications and postings. If a company wishes to pursue a system of monitoring then it should provide the necessary tools to take action if necessary – rather than imposing a duty which could be time consuming and difficult to sustain.

For more information on dealing with the media, see chapter 12 of The Nautical Institute's publication: *Casualty Management Guidelines*.

Contributors

Antonio Roberto M Abaya MD

Dr Abaya is Medical Director of Health Metrics Inc, a diagnostic centre for pre-employment medical examinations, in the Philippines. He has over 10 years experience of dealing with Filipino seafarers and is the first Filipino doctor with a master's degree in maritime medicine. After graduating from the University of the Philippines College of Medicine he spent the next eight years training in cardiovascular surgery in Spain, England and Belgium.

Chris Adams AFNI

Chris is a Partner and Director of the companies responsible for the management of the Steamship Mutual P&I Clubs (London and Bermuda) and currently Head of European Syndicate with responsibility for the business of all the Club's members domiciled in Europe, and Head of Loss Prevention with responsibility for all of the Club's loss prevention materials and initiatives. He previously served at sea as a navigating officer with Ellerman City Liners on general cargo, container and ro-ro ships.

Lynda Bailey

Lynda is a BACP accredited psychotherapist. She worked for CRUSE Bereavement Care for nine years as a counsellor, supervisor and trainer. She now runs Medra Counselling Services in North Wales, providing staff counselling to local authorities and businesses, as well as providing counselling to private clients.

Captain Trevor Bailey FNI

Trevor is currently Master of the luxury cruise vessel *Hebridean Princess* and a Younger Brother of Trinity House. He is also Chairman of The Nautical Institute's Technical Committee and a member of the Institute's Executive Board. His seagoing career started with BP Tanker Co, from where he moved to Sealink British Ferries becoming the first Training Master on board HSS *Stena Explorer*.

Contributors
The Nautical Institute on Command

Captain Sanjay Bhasin LLM MNI

Sanjay is a Director of Solis Marine Consultants in London. He is a highly experienced Master Mariner, with over 18 years' experience at sea on bulk, general, container and reefer vessels trading worldwide. He was in command for six years before coming ashore in 1999. Since then he has worked with P&I correspondents in South Africa and been responsible for implementing loss prevention policies for a leading cargo insurer.

Captain Sean Bolt BA MPS FCILT

Sean is currently Harbour Master and an unrestricted pilot at the Port of Albany, Western Australia. His seagoing experience included cargo ships, tankers and ro-ro vessels. He has held the position of CEO of a port authority, of a shipping company and of a stevedoring and marshalling company, and was a licensed pilot at the New Zealand ports of Tauranga and Westport, where he was also the Harbour Master.

Captain Richard Brough OBE BA AFNI

Richard is a Director of ICHCA, which represents cargo handlers and port operators around the world, and runs his own port and logistics consultancy. He has over 40 years experience in both the marine and shore-based sides of logistics, including 20 years at sea serving on all types of vessels carrying many different types of cargo. After coming ashore, he served in various roles in stevedoring management and port captaincy.

Lucy Budd BA

Lucy Budd is the editor of *Seaways*, the journal of The Nautical Institute, and managing editor of *The Navigator*. She joined the maritime industry in 2002, and was previously editor of *The Baltic* and *World Bunkering*. She has contributed to a number of publications as a freelance, including *Fairplay* and *Seatrade*.

Contributors

Captain Sarabjit Butalia MSc (WMU) FNI

Sarabjit is currently involved in maritime training with various Institutes in India, and undertakes training related assignments with V.Ships and DNV-GL Singapore. He sailed as Master on all types of dry cargo ships including containers and bulk carriers before joining V.Ships' Singapore office as Training Manager in 2005 where his primary responsibility was system implementation and developing new training modules.

Jillian Carson-Jackson MEd FNI AFRIN

Jillian has held the role of Vessel Tracking and Pilotage Services Manager with the Australian Maritime Safety Authority since 2006. Her experience afloat with the Canadian Coast Guard included ice-breaking, search and rescue and buoy-tending. She has taught at the Canadian Coast Guard College and worked with IALA as a technical coordination manager.

Captain Nicholas Cooper FNI MNM

Nicholas is currently a consultant with Cwaves, drawing on 35 years of seagoing experience of which 25 were in command on container ships and bulk carriers. He also has wide experience of ship and cargo superintendency and surveying and has implemented and audited quality and safety management systems. He is a Past President of The Nautical Institute.

Graham P Cowling ExC FNI FICS

Graham is Operations Manager for Marlow Navigation, Cyprus. He began his seagoing career in 1978 as a deck cadet with P&O Steam Navigation Company before moving to the Kuwait Oil Tanker Company. Ashore, he has worked in various ship management roles including fleet manager, technical director and operations manager with owners and managers based in Cyprus, Hong Kong and London. He is a member of The Nautical Institute's Council and Chairman of the Cyprus Branch.

Contributors
The Nautical Institute on Command

Kenneth W Crawford

Kenny currently works for Maritime New Zealand (MNZ) as the Manager for Navigation, Environment and International Operations and is also the Chair of the Technical Working Group of the Tokyo MOU. He served as chief engineer on ferries and anchor-handling vessels before coming ashore to work as a marine surveyor. Kenny is an Incident Controller for MNZ.

Captain Ozan Dermen MNI MSc BSc

Ozan has been a Master on Aframax and Suezmax tankers since 2010 and also teaches tanker and simulated electronic navigation courses at Holland College Marine Training Center in Canada. His experience includes acting as a technical consultant for a marine insurance brokerage company. Work for his master's degree at the University of New South Wales included adaptation of HAZOP risk analysis to cargo ship fires.

Christine Dickinson

Christine is a New Zealander who has worked as a personal assistant in the private and public sector in the UK and New Zealand. She has a private pilot's licence and has competed in an air race around New Zealand.

Captain Paul Drouin AFNI

Paul is currently the Principal of SafeShip.ca, a marine consulting firm specialising in marine investigations, safety culture and pilotage issues, and the editor of The Nautical Institute's Mariners' Alerting and Reporting Scheme (MARS). He spent 20 years as officer and Master with the Canadian Coast Guard and over a decade as a marine accident investigator with the Transportation Safety Board of Canada.

Contributors

Captain Neil Forde MNI

Neil is currently a nautical surveyor at the Marine Survey Office Ireland and the coordinating officer for Port State Control in the Cork regional office. He is also the responsible person for the competent authority for the IMDG Code and the national expert on hazardous cargoes. Neil obtained his first command in 1979 and remained as Master for the next 23 years with several companies trading worldwide on container ships.

Captain Pushkar Gadam MBA

Pushkar is a Risk Management and Loss Prevention Manager with DGS Marine Management Services, a fixed premium P&I insurance provider. He is a Master Mariner with more than a decade of seagoing experience, having served on bulk carriers and oil and LPG tankers. He has carried out numerous helicopter-ship operations while entering and leaving ports.

Michelle Rita Grech BMechEng MSc PhD CEng MNI

Michelle heads the Human Factors section within the Australian Maritime Safety Authority and current projects include supporting the development of human-centred design and fatigue management. A chartered engineer with over 20 years experience in the maritime domain, she has worked as a shipyard commissioning engineer, marine surveyor and maritime human factors researcher.

Peter Hamer MNI FRINA Master Mariner

Peter is currently the Area Manager for DNV-GL Maritime in Africa, responsible for offices in six countries, operations in 29 countries and 32 surveyors. He joined DNV in 1991 after 14 years at sea and gaining his Master's ticket and completing a BEng in naval architecture. His experience with DNV included acing as project manager on cable layer and drillship conversions.

Contributors
The Nautical Institute on Command

Captain Ghulam Hussain MBA FICS FNI

Ghulam joined The Nautical Institute as Accreditation Manager in September 2014 and is responsible for accreditation of the training providers for The Nautical Institute's Dynamic Positioning Operator training scheme. He is a Master Mariner, having sailed for 14 years, and since coming ashore he has lived and worked in Hong Kong and Bangladesh in a variety of roles in chartering, shipbroking and ship management.

Steven Jones BSc (Hons) MSc MNI DipMarSur FRSA

Steven is currently a consultant, specialising in security and safety planning, and the author of The Nautical Institute's maritime security suite. He spent a decade as a navigation officer and since coming ashore has worked across the maritime industry, within shipping companies, insurers, publishers and professional bodies. He was involved in the development of the Security Association for the Maritime Industry and founder of the International Dynamic Positioning Operators Association.

Captain Robert Kieran LLM FCIArb FNI

Bob is a marine consultant and qualified pilot and currently chairman of the Association of Maritime Pilots Ireland. After gaining his Master's certificate and a master's degree in marine law, he worked with Admiralty solicitors and for a major P&I club in its legal and liability claims departments. He is a Fellow of the Chartered Institute of Arbitration and has also worked as a surveyor and ISM training consultant.

Tapan Kumar

Tapan is currently the General Manager of Seachef, the maritime catering division of Bernhard Schulte Ship Management. He has been actively involved in developing training courses for the catering staff in Mumbai, Manila, China and Cyprus and has introduced a Food Safety Management System based on MLC regulations on board Seachef served vessels.

Contributors

Dr Captain François Laffoucrière AFNI ACIArb

François is currently a marine consultant, a listed marine surveyor, a maritime arbitrator and a trainee advocate. He is a Master Mariner and a chief engineer, having taken command of crane barges, drill-ships and semi-submersibles, and was a pilot at Le Havre from 2000 to 2013. He has a PhD in law from the Sorbonne and acted as judge for the Commercial Court of Le Havre for several years.

Rear Admiral Nick Lambert CMarTech FIMarEST

A Master Mariner, Nick concluded a long naval operational career as the UK National Hydrographer in December 2012. He advises on a wide range of maritime issues including the importance of hydrography for maritime economies, evolution of eNavigation and GNSS vulnerability, near or real time situational awareness, human factors and maritime education and training.

Captain André L Le Goubin MNM MA FNI

André is currently employed as a Mooring Master in the lightering trade, undertaking ship-to-ship transfers in the Gulf of Mexico. He has recently started his own company, DNA Marine, where he specialises in providing expert witness services. His experience includes command of high-speed ferries (hydrofoils, mono- and multi-hull ro-ro/passenger vessels), working as a pilot for the London Pilotage Service and acting as a marine consultant.

Ian MacLean MBA MNI

Ian is a partner at the international law firm Hill Dickinson. A Master Mariner with 12 years seagoing experience, he has since been involved in casualty investigations for over 20 years and advises on contractual and third party disputes arising from casualties, salvage, unsafe ports, MARPOL, hull and machinery claims and ship management. He was awarded a Nautical Institute scholarship in 1992 to complete an MBA.

Contributors
The Nautical Institute on Command

Iain Macleod

Iain retired recently from Northern Marine, part of the Stena Group, as the company's senior manager responsible for all seafarer and shore staff dispute resolution. He was responsible for the timely implementation of the Maritime Labour Convention for the whole Northern Marine fleet of over 100 vessels, covering at least 16 flag states, and has considerable experience in human resources performance management.

Captain Wendy Maughan FNI K(DK)

Wendy is a Master at Wightlinks Lymington to Yarmouth ferry service and a Younger Brother of Trinity House. She was awarded Fellowship of The Nautical Institute for her services to the industry and is a Knight of the Order of the Dannebrog for her continuing service to the Danish Embassy as Southampton Vice-Consul. Her onshore experience includes management roles in stevedoring companies, including 13 years as General Manager of Berkeley Handling.

Captain Nick Nash FRGS FRIN FNI YB

Nick is a serving Master with Princess Cruises, part of the Carnival Group, and has over 33 years experience at sea. He has served on general cargo ships, reefers, tankers, container ships, North Atlantic ro-ros and in the RFA. He joined P&O/Princess Cruises in 1989 and was promoted to Staff Captain in 1997 and Captain in 2002. He has taught and is a consultant at Carnival Group's marine simulator training facility CSMART and a Vice-President of The Nautical Institute.

James Parkhouse BEng (Hons) CEng MRINA

James is currently Assistant Director for the Bahamas Maritime Authority where he is involved in drafting and implementing flag state policy for Port State Control, lifesaving appliances and oversight of Recognised Organisations. He is a naval architect and chartered engineer and has previously worked in a variety of regulatory roles, including as a class surveyor and in the unique classification and regulatory regimes applied to naval ships.

Contributors

Eric Patten Captain USN (Retd)

Eric is the CEO and President of Ocean Aero, an unmanned maritime systems company developing an autonomous unmanned surface and subsurface ocean observation platform. Before joining Ocean Aero, he was Director, Defense and Intelligence Global Solutions for the geospatial technology company Esri. His 25-year career in the US Navy included command of Helicopter Anti-Submarine Squadron Five One in Japan and leading combat missions as a naval aviator.

Reverend Canon Ken Peters Dip Th MBA MA FNI MIW RNR

Ken is currently the Director of Justice and Public Affairs for The Mission to Seafarers. His previous role was Director of Justice and Welfare. He is an experienced Port Chaplain for The Mission to Seafarers, serving in Europe and Japan from 1980 to 1994.

Captain Sriram Rajagopal MA (London) MICS

Sriram is currently a maritime consultant. He started his career with Barber Ship Management (now Wilhelmsen) where he sailed for 17 years as an officer and Master on board car and bulk carriers, ro-ros, container and general cargo ships, and crude oil tankers. On coming ashore he worked for Anglo Eastern Ship Management covering QHSE, internal audit and training. He was awarded the Medite prize by the Institute of Chartered Shipbrokers in 2004.

Captain James Robinson DSM FNI Irish Navy (Retd)

James is a past President of the Nautical Institute and is currently Chairman of the Executive Board. Before his retirement in 2009 he had been a mariner for 42 years, having spent 36 years in the Irish Navy and six years with Irish Shipping. Service with Irish Shipping included a cadetship and periods as third and second officer. He retired as Officer Commanding, Naval Operations Command and Second in Command Naval Service.

Contributors
The Nautical Institute on Command

Captain Alexander Sagaydak FNI MIMarEST

Alexander is currently the CEO of the Ukrainian crewing company Olvia Maritime and since 2000 has also worked for the IMO as an expert, lecturer and trainer covering ballast water management, anti-fouling systems and energy efficiency. He has been in command of multi-purpose, training and passenger ships and is also a part-time lecturer at the Odessa National Maritime Academy.

Kevin Slade

Kevin is currently Director of Northern Marine Management, a wholly owned Stena company, and a member of the Merchant Navy Training Board. He served 21 years at sea, nine years as Master before moving ashore as Marine Manager with Sea Containers in 1987. He moved to personnel management with Stena Line in 1991 and joined Northern Marine Management in 1997, becoming Personnel Director in 2003.

Captain David (Duke) Snider FNI FRGS

Duke is the CEO and Principal Consultant of Martech Polar Consulting, which provides global ice navigation services and support for polar shipping. He is a Master Mariner and spent 29 years at sea operating vessels in various ice regimes, retiring from Canadian Coast Guard service as Regional Director Fleet Western Region in 2012. He has served on naval, commercial and Coast Guard vessels and is Senior Vice-President of The Nautical Institute.

Captain Richard Springthorpe BSc MNI

Richard is currently Operations Manager for Stolt's Chemical Terminal and Tanker division in New Orleans. He first went to sea in 1991, working for Stolt-Nielsen, and sailed until 1996 as an officer. After gaining a BSc in marine technology he worked for Northern Marine Management sailing on VLCCs and chemical tankers as chief officer and then Master. In 2008 he joined the international law firm Ince & Co where he handled maritime casualty cases and disputes.

Contributors

Captain Kuba Szymanski FNI

Kuba is the Secretary General of InterManager, the trade association for ship managers, and lectures on the superintendency courses at the International Business School, Isle of Man. He began his sea career in 1985 and has sailed on gas carriers and chemical and product tankers, reaching his first command as Master in 1999. On coming ashore, he worked as a marine superintendent.

Walter Vervloesem FNI

Walter founded the IMCS Training Academy in 2014, having been involved in the expansion of the IMCS Group to 19 branch offices worldwide and becoming its Chairman in 2000. He became a marine surveyor on leaving the sea in 1988 with the rank of chief officer on short-sea trade vessels. He pioneered the use of ultrasound for testing the weathertight integrity of hatch covers and trains people in its use.

Captain Alex van Wijngaarden AFNI

Alex was appointed Harbour Master for Marlborough in New Zealand in 1994, a position held to date. His maritime career spans 45 years, both at sea and ashore, and he has served on tankers, general cargo, passenger and container ships, and reefers. In 1978 he gained his first command, of a small tanker in the Persian Gulf. He is a Vice-President outside Europe for the International Harbour Masters' Association.

Alison Williams BA (Hons) FCIPD

Alison is currently European HR Manager for Svitzer, leading a team of HR business partners, crewing managers and training professionals. She is also a trained career coach and mentor. For much of her career, she has been in a generalist HR role but has also specialised in industrial and employee relations and managing the HR aspects of large business transformation programmes.

144 | Also available

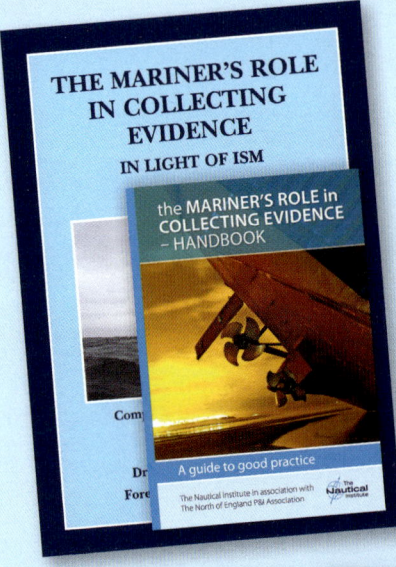
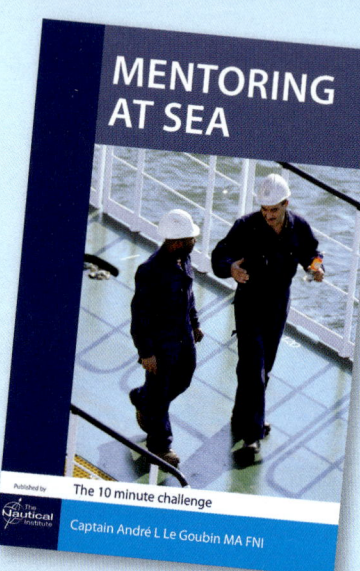
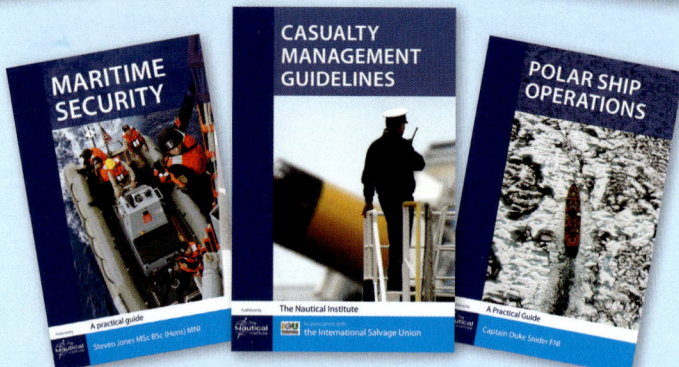